# CLASSIFICATION AND IDENTIFICATION OF FRESHWATER FISHES

# About the Authors

Prof. Hiware Chandrashekhar J. is presently on deputation and working as a Director, Sericulture, Maharashtra State, but he is Professor at Department of Zoology, Dr. Babasaheb Marathwada University, Aurangabad (M.S.) India. He has completed his M.Sc with distinction and Ph.D from same University. He has also completed P.G.D.S. from C.S.R & T.I., Mysore. He has 25 years teaching experience. He published 114 research articles through different journals of national and international repute. Under his able guidance 19 students and 01 student completed Ph.D. & M.Phil Degree respectively. He is actively engage in teaching, research in Icthyo-Parasitology and Sericulture, scientific and social activities.

Dr. Pawar Rajkumar T. is presently working as a Head, Department of Zoology, M.S.P. Mandal's Majalgaon Arts, Science and Commerce College, Majalgaon, Dist. Beed (M.S.). He has completed his M.Sc. and Ph.D. Degree from Dr. Babasaheb Ambedkar Marathwada University, Aurangabad. He had 8 years research experience and four year teaching experience. Eleven Research articles are there in his credit in reputed National and International Journals.

Dr. Gaikwad Jaiprakash M. is presently working as Associate Professor and Head Department of Fisheries, Shri Shivaji College Parbhani,(MS),India 431401.He has completed his M. Sc. and Ph.D. from Dr. Babasaheb Ambedkar Marathwada University Aurangabad. He has 24 years of teaching experience at UG Level and 21 years at PG Level. He has published 25 research papers in different National and International reputed Journals.

Dr. (Mrs.) Sonawane Smita R. is Professor at Department of Zoology, Dr. Babasaheb Marathwada University, Aurangabad- 431004 (M.S.) India. She has completed her M.Sc. with distinction and Ph.D. from same University. She has 32 years teaching experience. She has published 45 research articles through different journals of national and international repute.

# CLASSIFICATION AND IDENTIFICATION OF FRESHWATER FISHES

Prof. Hiware Chandrashekhar J.
Dr. Pawar Rajkumar T.
Dr. Gaikwad Jaiprakash M.
Dr. (Mrs.) Sonawane Smita R.

2015
Daya  Publishing House®
*A Division of*
Astral International (P) Ltd
New Delhi 110 002

Published by                          : **Daya Publishing House®**
                                        A Division of
                                        **Astral International Pvt. Ltd.**
                                        – ISO 9001:2008 Certified Company –
                                        4760-61/23, Ansari Road, Darya Ganj
                                        New Delhi-110 002
                                        Ph. 011-43549197, 23278134
                                        E-mail: info@astralint.com
                                        Website: www.astralint.com

Laser Typesetting                     : Classic Computer Services, Delhi - 110 035

Printed at                            : Replika Press Pvt. Ltd.

PRINTED IN INDIA

# Foreword

My sojourn to Babasaheb Ambedkar Marathwada University, Aurangabad from time-to-time gave me an opportunity to meet Dr. Chandrashekhar Hiware, Professor of Zoology in the Department of Zoology, and to study his work on Cestode parasites of fishes found in water bodies of the Marathwada region. During his investigation, he collected fishes and arranged systematically to identify them this led him to write this book on "Classification and Identification of Freshwater Fishes". On the basis of morphological variations and quantitative measurements of morphometric characters, he prepared artificial key to identify them species wise, so that it will be useful information for conservation of biodiversity of this region, hobbyists, traders of aquarium fishes, fishery department and aquaculture entrepneurists and more specifically to local fishermen and farmers. This book should be recommended as a course book for graduate and postgraduate students of Marathwada University. It will be easy to the Biodiversity Board of Maharashtra State to prepare Peoples' Biodiversity Register on freshwater fishes based on this information. At the same time this useful information will help the Regional Team of Zoological Survey of India to prepare update records on freshwater fishes from time-to-time. This book is of immense benefit to research workers in the field of physiology and biochemistry of fishes by identifying exact species name and location of its habitat. The Bird Watchers would get proper clue to certain species of waterfowls based on systematic information on fishes given in this book and thus would be useful for eco-tourists who would visit this region.

**Dr. D.K. Belsare**
Ph.D., D.Sc.F.N.A.Sc, F.Z.S.I.,
Former Dean, Life Science Faculty,
Barkatullah University, Bhopal

# Foreword

The authors of the present reference book, Prof.C.J. Hiware *et. al.* have not only accomplished a pioneering but also commendable work in the field of fish biodiversity of Marathwada Region of Maharashtra State. This book encapsulates diverse invaluable information pertaining to as on many as 76 species of fishes, belonging to 43 genera, under 20 families, inhabiting freshwater-bodies like wetlands, rivers and lakes, etc. of Marathwada Region.

This book provides detailed information on fish taxonomy and faunal diversity of the region of paramount zoo-geographical significance. The taxonomy of fishes has been updated in the light of latest systematic changes. The morphology of the fish species, identification keys, diagrams and figures have been nicely highlighted. The book will serve as an indispensable tool for identification of fish-fauna. Apart from this, the baseline data generated will undoubtedly cater to the needs of stakeholders in the area of fishery sector, with sustainable utilization, conservation and management.

This invaluable publication, prepared painstakingly, has great significance in the bio wealth knowledge of Maharashtra State of Indian sub-continent and will satisfy the growing needs of Students, Researchers, Scientists, Environmentalists and Bio-resource Managers, etc.

**Prof. (Dr.) R.C. Bhagat**
(Former, Head, Deptt. of Zoology,
& Dean, Faculty of Sciences,
University of Kashmir (J& K), India)

# Foreword

The authors of the present reference book entitled as "Classification and Identification of Freshwater Fishes" have very nicely dealt with the general aspects, morphological variations, measurements and counts used for the diagnosis of the fishes and also explained the systematics of fishes of Marathwada region. The contribution made by the authors is a milestone and will help as a mentor for the researchers who are actively engaged in the area of freshwater fish research.

They have not only taken a pioneering step but also done commendable work in the field of fish biodiversity of Marathwada Region of Maharashtra State. This book encapsulates diverse invaluable information pertaining to different fish species inhabiting freshwater-bodies like wetlands, rivers and lakes, etc. of Marathwada Region of Maharashtra state.

As there is non-availability of any good book of freshwater fish taxonomy of this region, this book will be very helpful and the authors have attempted very hard task of bringing out the useful monograph. The book will serve as an indispensable tool for identification of fish-fauna. I am sure this book will be immensely useful to the college and university teachers, researchers, students, fishery scientist engaged in studies on freshwater ichthyology. I congratulate authors for bringing out this valuable piece-of-work on freshwater fishes of Marathwada region and wish them all the success.

**Dr. Ashok Mohekar,**
Principal, S.M.D.M. Mahavidyalaya,
Kallam and Dean, Faculty of Science,
Dr. B.A.M. University, Aurangabad
(Maharashtra State)

# Foreword

The fishes are most widely distributed vertebrates in the vast water areas of the earth. The present title "Classification and Identification of Freshwater Fishes" written by Dr. Hiware Chandrasekhar *et. al.* deals with the general aspect, morphological variation, measurements and counts used for the diagnosis and systematics of fishes of Marathwada region. Contribution made by Dr. Hiware is a milestone and mentor for the researchers who are actively engaged in the above area of research. It was hobby of Dr. Hiware and his co-authors to visit different freshwater-bodies of the region to study their ichthyofauna.

It is very heartening to note that inspite of his busy schedule as a professor of a university and director of the end of his long carrier Dr. Hiware kept his interest in fishes alive and always showed interest to identify a fish from their region. The fish taxonomy is very essential so that we very correctly identify, conserve and properly manage the fish diversity of freshwater of Marathwada region. Morphometric methods form the backbone of taxonomy but after methods based on new technique of molecular biology and biotechnology are also emerging these days.

After the non-availability of any good book of freshwater fish taxonomy of this region, authors book will prove very helpful. Their experiences as a teacher and guiding research in fish taxonomy have gone a long work in bringing out the useful monograph. I am sure this book will be immensely useful to the college and university teachers, researchers, students, fishery scientist engaged in studies on fresh water ichthyology. I congratulate Dr. Hiware for bringing out this valuable piece-of-work on Classification and Identification of Freshwater Fishes.

**Prof. (Dr.) D.R. Khanna,**
Former Head,
Department of Zoology & Environmental Science,
Gurukula Kangri Vishwavidyalaya,
Haridwar – 249 404 (India)

# Preface

Freshwater resources are always under increasing pressure throughout the world. The human population is always under accelerated growth rate and the per capita water demand for a variety of uses is also high and increasing. Freshwater resources are becoming forever more important because of its many fold utility. As the freshwater ecosystems are dynamic and always exposed to different situations. The fishes are most important component inhabiting in such resources along with the other organisms and among the most threatened unit worldwide with a large percentage of species assigned to some level of danger. In many parts of our country the freshwater fisheries providing gainful employment and is an important source of protein rich food for local populations. It has commercial and recreational value also. The Freshwater fishes are having especial importance among the aquatic biota as they serve the dual roles as major sources of food protein and also its important role in the health and vitality of freshwater ecosystems.

The process of conception of this book was started from 2006 and after a long time gap it has came in existence. At that time, University Grants Commission, New Delhi had funded 3-years project on "Diversity, Population dynamics and histopathological studies of Cestode parasites from Freshwater fishes of Marathwada region, Maharashtra State, India" and it was carried out at the Department of Zoology, Dr. Babasaheb Ambedkar Marathwada University, Aurangabad, Maharashtra State, India. From the start of the project it was thought to have the book on the Freshwater Fishes of this region and it was organized, accordingly arranged to collect the total relevant data. One of the objectives in the project was to study the diversity of Piscean fauna from Marathwada region and accordingly by selecting some important freshwater resources and the localities the work had been carried out. It is the sincere attempt made to record the biological diversity of piscian fauna from the Marathwada region which differ greatly in their taxonomic composition and species richness.

The content of this book presents the general aspects of the fish identification and the systematics of fishes collected from different localities of Marathwada region of Maharashtra State. The book is organized in four main chapters.

The first chapter describes the main components of General aspects, classification and key for the identification of fishes. The second chapter deals with the morphological variations. The third chapter presents various measurements and counts used for the diagnosis. Finally, the fourth chapter is devoted to the Systematics of fishes.

We hope this book will be beneficial to the students seeking knowledge in fishery science subject at degree, post-graduate and research level. It provides the baseline data for the researchers and the personnel's interested in the fishery sector in Marathwada region.

Authors acknowledges with gratitude the support in various levels rendered by different peoples during collection of data while preparation of this book. Authors highly acknowledge with obligation the teachers, friends and the people, who deliberately contributed their help directly or indirectly by presenting their interest in greatly enhancing the contents of this book.

It is anticipated that the relevance of the book will be such that anyone with interests in fish and fisheries will definitely satisfy their need. Though also Authors are thankful to them who pass the constructive suggestions, comments, criticism etc. and are highly appreciated; those will be considered as an asset for the improvement of the standard of this book.

**Prof. Hiware Chandrashekhar J.**
**Dr. Pawar Rajkumar T.**
**Dr. Gaikwad Jaiprakash M.**
**Dr. (Mrs.). Sonawane Smita R.**

# Acknowledgements

Due to the co-operation, best wishes and continuous inspirations from many peoples we could complete this book. We would like to express our gratitude to all those people who has provided direct or indirect help.

We would like to thanks University Grants Commission, New Delhi for their generous funding grants for 3-years on project on "Diversity, Population dynamics and histopathological studies of cestode parasites from Freshwater fishes of Marathwada region, Maharashtra State, India" through which we could convert the collected data information opportunity into book form.

We would sincerely thanks to head Department of Zoology, Dr. Babasaheb Ambedkar Marathwada University, Aurangabad, Maharashtra State, India, for providing the laboratory facilities for this work.

The authors would like to thank Anil Mittal, Director, Astral International (P) Ltd., New Delhi for helping in the process of selection and printing this book and bringing out this useful publication.

Last and not least we beg forgiveness of all those who have been with us over the course of the years and whose names we have failed to mention here but we will feel very happy to remain forever in their debt without saying thanks to them.

**Prof. Hiware Chandrashekhar J.**
**Dr. Pawar Rajkumar T.**
**Dr. Gaikwad Jaiprakash M.**
**Dr. (Mrs.). Sonawane Smita R.**

# Acknowledgements

Due to the co-operation, best wishes and continuous inspirations from many peoples we could complete this book. We would like to express our gratitude to all those people who has provided direct or indirect help.

We would like to thanks University Grants Commission, New Delhi for their generous funding grants for 3-years on project on "Diversity, Population dynamics and histopathological studies of cestode parasites from Freshwater fishes of Marathwada region, Maharashtra State, India" through which we could convert the collected data information opportunity into book form.

We would sincerely thanks to head Department of Zoology, Dr. Babasaheb Ambedkar Marathwada University, Aurangabad, Maharashtra State, India, for providing the laboratory facilities for this work.

The authors would like to thank Anil Mittal, Director, Astral International (P) Ltd., New Delhi for helping in the process of selection and printing this book and bringing out this useful publication.

Last and not least we beg forgiveness of all those who have been with us over the course of the years and whose names we have failed to mention here but we will feel very happy to remain forever in their debt without saying thanks to them.

**Prof. Hiware Chandrashekhar J.**
**Dr. Pawar Rajkumar T.**
**Dr. Gaikwad Jaiprakash M.**
**Dr. (Mrs.). Sonawane Smita R.**

# Introduction

India has a large part of its land mass surrounded by water. It boasts of a long coastline of about 7000 km and has 40 major rivers, not to mention numerous stagnant water bodies like lakes, tanks, and reservoirs. Almost all these water-bodies have a variety of fish species living in them. Hence it is not surprising that a large section of the Indian population depends on fishes (both freshwater and marine) for its food and livelihood. We have as many as 2500 species of fishes of which about 930 (40% of the total!) are freshwater inhabitants. Though there have been some studies on the classification of fishes, the first modern, and scientific method for classifying fishes of the Indian region was the colossal work by Hamilton-Buchanan on the fishes of the Ganges in 1822. A number of workers in the 19th century, like J. McClelland, Col. W. Sykes, T. C. Jerdon, Blyth and Francis Day contributed to the study of fish taxonomy. Of these, Francis Day's (1875–1878) Fishes of India (London) and Fauna of British India (Vol. I and II) are very relevant and widely referred to even today. The foundations of modern-day classification of Indian fishes were laid by the studies of these pioneers. In the 20th century, the studies on the taxonomy of fishes in India were carried on to a large extent by scientists working at the Zoological Survey of India (ZSI). One of the greatest ichthyologists of India in the 20th century was Sunder Lal Hora (1920–1955), who paved the way for a number of other scientists at the ZSI, including K. C. Jayaram to continue their studies on the ichthyofauna of the region. A number of publications have come up since the days of Hora, which deal with the taxonomy of the fishes region-wise and family-wise. With as many as 930 species belonging to about 70 families and 280 genera of freshwater fishes found in the Indian region, it would be an immense amount of work to put this information into one comprehensive book. In 1981, K. C. Jayaram first published his Handbook on the Freshwater Fishes of India, Pakistan, Bangladesh, Burma and Sri Lanka. It was an instant success and a welcome addition to the available literature since it was the first comprehensive and handy reference volume, which covered almost all freshwater fish-fauna in this region. Talwar and Jhingran (1991) published two very good volumes on Inland Fishes of India and Adjacent Countries (Oxford and

IBH Publishing Co Pvt Ltd, New Delhi). Subsequently, there have been changes in the classification and renaming of a number of taxa and there was a need for an update of these works.

The Marathwada region is under intense anthropomorphic pressure due to population growth, pollution, and continued modification of land for agriculture and forestry. Consequently, there is an urgency to study organisms in this region. In addition, the isolation of the fauna makes it quite unique. The fish communities that have evolved in this region occur nowhere else on Earth. Understanding how species and communities have responded can increase our understanding of how organisms respond to major environmental disturbances.

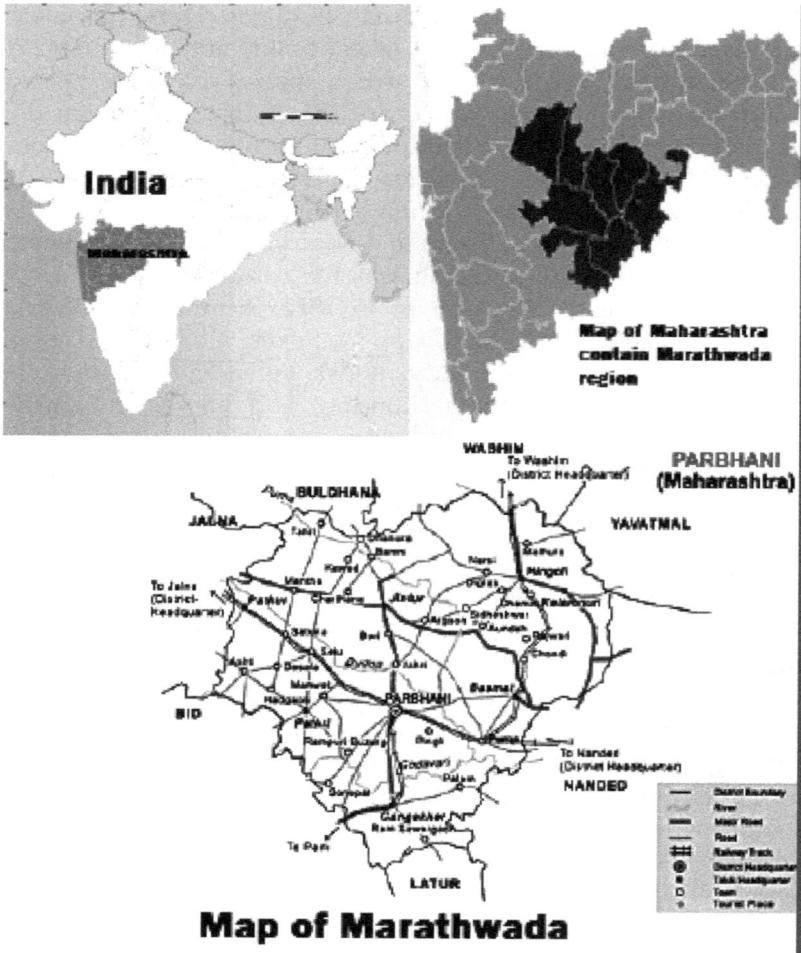

Map showing the study area of Marathwada Region, Maharashtra, India.

Today, everybody agrees that freshwater biological diversity is threatened, so, we are particularly interested in the expanding knowledge of freshwater fishes, and of their taxonomic diversity as the very first step. The present revision is prepared within the Major Research project "Diversity population Dynamics and Histopathological studies of cestode parasites of Freshwater fishes from Marathwada region". The main objective was to synthesize data on the diversity of fishes living in lakes, rivers, streams and reservoirs of Marathwada region.

A total of 76 species, belonging to 43 genera in 20 families have been included in this book. The freshwater fish-fauna of Marathwada, Maharashtra mostly occupies the wetlands, rivers and lakes. The numerically dominant families are: Cyprinidae (at least 36 species in 15 genera), Bagaridae (07 species in 02 genera), Channidae (04 species in 1 genera), Cobitidae (04 species in 02 genera), Schilbeidae (03 species in 03 genera), Poecilidae (03 species in 02 genera), Notopteridae (02 species in 01 genera), Claridae (02 species in 01 genera), Siluridae (02 species in 02 genera), Mugilidae (02 species in 02 genera), Ambassidae (02 species in 02 genera), Mastacembelidae (02 species in 01 genera), and Heteropneustidae, Sisoridae, Belonidae, Anabantidae, Anguilldae, Gobiidae, Cichlidae, Nandidae, (01 species in 01 genera).

The book needs of a wide audience, ranging from graduate and undergraduate students and also researchers who need a manual for the identification and taxonomy of fishes. This book has information on work done in recent years which have led to renaming and reclassification of some taxa. The arrangement of keys has been done in a very 'user friendly' manner such that one need not have too much prior experience with fish identification to be able to use it. The features used for classification are mostly external morphological ones (body shape, length, depth, presence or absence of spines, barbels, fins, Mouth, snout, jaws, teeth, scales, colour of the species, etc.) such that tedious dissections can be avoided as far as possible. At the same time the latest classification criterion has been used in accordance with international conventions of classification of fish taxa.

A beautiful description on how to take meristic measurements and accounts of various features of the fish specimen that can be recorded follow this. And each of these has in addition to the description very clear diagrams, which make understanding of each feature in the fish very easy. Here too, at places some descriptions have been supplemented with diagrams of the features. A systematic index is provided at the very beginning of the book where all the genera described in the book have been arranged according to the conventional method of classification into superclass, class, subclass, division, order, suborder, family and genus.

A key for identifying each order follows this. Once that is done, a more elaborate description of each order has been provided along with the families covered within each order. Again within each family, a set of diagnostic features for that family is given along with the range of distribution and the number of genera belonging to that family. Elaborate descriptions of each genus are given, including information on the naming of the genus, its diagnosis, distribution and the number of species of the given genus. At places the author has also given additional information on the importance of a genus in terms of commercial value and spawning habits and also peculiar feeding habits which give interesting additional information about the genus. A description of each genus is concluded with a list of species belonging to that genus along with its range of distribution. At least one figure of a representative species of the genus is given and this is followed by a key to the species. The most impressive feature of the book is the excellent quality of these diagrams and figures, which makes identification so much easier. It would be too much to ask for to have further descriptions of each species and still want it to remain an easy reference single volume handbook! But an elaborate description of each species is hardly a requirement for most researchers and fieldworkers who want an identification of their fresh specimens in the field itself. A useful feature, has added for biologist, *i.e.* fin formula of a species.

There is little published on it and what exists is broadly scattered and often difficult to access. It would be very useful for further reading if any researcher needs additional information on specific taxa or topics. This book is to provide easy access to present knowledge of the freshwater fishes of Marathwada region and is must for every students of ichthyology to possess as a ready reference to fish identification.

# Contents

Division: Cyprini
Order II : Cypriniformes
Sub-order : Cyprinoidei
Family 2 : Cyprinidae
Subfamily : Abramidinae

Subfamily : Cyprininae

Subfamily : Rasborinae

Subfamily : Poeciliinae

# 1

# GENERAL ASPECTS OF FISHES

## 1.1 CHARACTERISTICS OF FISHES

Fishes exhibit a number of specialized features, which distinguish them from the other groups of animals very easily. The fishes are purely aquatic and cold-blooded vertebrates that breathe by means of gills. They are devoid of a typical lung and take their oxygen supply from water through gills. They live in almost every place where there is water. They are having fins, the appendages that help them in swimming and balancing their bodies. The fins are paired and unpaired supported by soft or spiny rays. Dorsal, Caudal and the anal fins are unpaired while the pectorals and the pelvic (ventral) are paired. The skin is usually covered with scales and is profusely coated with mucus. Some are without scale. The internal ears is present. Peculiar sense organs are found scattered on the head and on the lateral sides of the body to relate them to watery condition of current heat pressure and electromagnetic waves. They constitute economically a very important group of animals. Besides being used as food, fish liver oil is an important source of oil containing vitamins A and D, fish body oil is extensively used in soap industry and tanning. Fishes also yield fish meal, fish manure, is in glass and several other products of commercial importance. Most fishes are oviparous but some of them also are viviparous or ovoviviparous. In some of the fishes the parental care is also observed.

Thus the fish can be defined as a group of cold-blooded aquatic, gnathostomous vertebrates, which breathe by means of branchial gills and locomotion is carried by means of fins.

## 1.2. CLASSIFICATION

First scientific classification of the lower vertebrates was given by J. Muller (1844). He divided 'Pisces' into six Sub-Classes Dipnoi, Teleosti,

Ganoidei, Elasmobranchii, Marsipobranchii (Cyclostomi) and Leptocardii (Amphioxini). Agasslz (1857) separated lampreys and hagfishes into a separate class Myzonets, Pisces, Ganoidei, and Sellachii. Boulenger (1904) divided the Teleostean fishes into thirteen orders. Since then a number of systems of classification have been propounded [Gunther (1859-1870), Day (1878-1889), Regan (1906-1929), Jorden (1923), Goodrich (1930), Berg (1940), Grasse (1958), Romer (1959), Nikol'skii (1962) and Greenwood et al. (1966)]. Regan (1906-1929) had given an elaborate and widely accepted classification of fishes. Jorden (1923) divides fish like vertebrates into six classes. Among the more recent authors Goodrich (1930), Berg (1940), Grasse (1958), Romer (1959), Nikol'skii (1962) and Greenwood et al. (1966) have proposed a new scheme of classifying and teleostean fishes in detail. The system of classification by Berg (1940) is adopted here largely. According to him the Pisces or fishes are placed under the following scheme of classification.

Phylum – Vertebrata

Sub-phylum – Craniata

Super-class – Gnathostomata

Series – Pisces

The series Pisces is divided in to six classes (Berg, 1940) as below:

1. Pterichthys

2. Coccostei

3. Acanthodii

4. Holocephali

5. Elasmobranchii

6. Dipnoi

7. Teleostomi

The class Teleostomi is further divided in to two sub-classes as Actinopterygii and Crossopterygii *i.e.* Ray fined fishes and lobed fin fishes respectively.

Out of the classes mentioned above, the first three classes (Pterichthyes, Coccostei and Acanthodii) are completely extinct and are collectively termed as Placoderms. The living forms are grouped under the four-remaining classes of the above scheme. The Elasmobranchii, Holocephali and Dipnoi are not considered in this text. In the present work, the species of Teleostean fishes found from Marathwada region of Maharashtra State, India are taken into account.

## 1.3 KEY FOR THE IDENTIFICATION OF FISHES

### Key to the Classes of Series Pisces

I. Endoskeleton cartilaginous, skull without cranial sutures gill pouch like and attached by their outer edge to the skin, whilst an intervening gill opening exits between each five pairs of lateral or ventral.

## Class Elasmobranchii

II. Endoskeleton bony, skull possessing cranial bones vertebrate completely separated and the posterior extremity of the vertebral column bony or having bony plates. Branchiostegeal rays present. A pair of lateral gill opening as a single ventral slit or non-confluent as two lateral slits with operculum.

## Class Teleostomi

*Characters*

The endoskeleton is bony. The primary upper and lower jaws are supplemented by the addition of membrane bones which from the secondary jaws and hence they are termed 'Teleostomi' or perfect mouthed fishes. The skull is usually hyostylic or amphistylic. Palatoquadrate is not fused with the Cranium. There are five gills arches and four gills on each side of the head. There is a pair of lateral gill opening confluent as a single ventral slit or non-confluent as a pair of lateral slits each with a gill cover, the operculum. Inter branchial septa are greatly reduced. Branchial lamellae are supported by a double row of branchial rays. The body is generally covered with ganoid, cycloid or ctenoid scales or nacked. Cosmoid or cycloid scales are rarely found. The cloaca and the claspers are absent. The air bladder is usually present.

The class Teleostomi is divided into two sub-classes:

(A) Crossopterygii (Lobe fin fishes)

(B) Actinopterygii (Ray fin fishes)

## Subclass Crossopterygii (Lobe fin fishes)

Paired fins with a scale covered lobe and supported by an endoskeleton consisting of a jointed median axis with radials on each side. The scales are Cosmoid or cycloid type. Internal nostrils are present. Dorsal fins are two in number. The jugal sensory canal traverses squamosal. This subclass has no Indian representative *e.g.*; Latimeria.

## Subclass Actinopterygii (Ray fin fishes)

Paired fins are without muscular lobes. The radials of the fins are not arranged biserally. Internal nostrils are absent. Scales are not of Cosmoid type. Squamosal and jugal sensory canals are absent.

Subclass Actinopterygii sometimes is divided into two groups Chondrostei and Holostei (Goodrich, 1909) he included in Holostei both, Holostei and Teleostei stensio (1932) divided it into three groups Viz. Chondrostei, Holostei and Teleostei. Regan (1929) divided the subclass Actinopterygii into two groups; Palaeopterygii (corresponding to Chondrostei) and Neopterygii (corresponding to Holostei and Teleostei). However the work of Stensio (1932) has shown that the Chondrostei gradually passes into Holostei and they can be separated from each other only artificially. Similarly, it is not possible to clearly differentiate Holostei from the Teleostei. Hence according to Berg (1940), if both the living and fossils forms are taken into consideration, the division of Actinopterygii into Chondrostei, Holostei and Teleostei becomes artificial. Berg has therefore abolished these three groups and divided the Actinopterygii into a series of orders. In the following text only seven orders are taken accounts.

Synopsis of the orders of the subclass Actinopterygii.

## KEY TO THE ORDERS OF SUBCLASS ACTINOPTERYGII

    I. Abdomen keeled and serrated.
    Barbels always absent.
    Head and body with small scales.
    Lateral line absent.........................................................CLUPEIFORMES

## Key to the Suborder of order CLUPEIFORMES

    I. Body neither elongate, nor narrow.
    Caudal fin is not bifurcate.
    Anal fin is continuous with, the caudal fins.......NOTOPTEROIDEI

## Key to the family of sub-order NOTOPTEROIDEI

    I. Body is moderately elongated and compressed upturned tail.
    Abdominal edge serrated anterior to the ventral fins.
    Branchiostegal rays from 3 to 9
    Head and body with small scales
    Barbels are absent
    Adipose fin is absent ....................................................NOTOPTERIDAE

## Key to the genera of family NOTOPTERIDAE

Caudal region is long and tapering
Snout is obtuse and convex, cleft of mouth is lateral
Teeth in jaws premaxillary, maxillary, vomer, palatine
Pterygoid and tongue bear small teeth
Pharyngeal teeth are absent
A single rayed dorsal fin and its originates for behind
The origin of anal
Anal is very extensive and united with the caudal
.............................................................................................. *NOTOPTERUS*

## Key to the species of genera NOTOPTERUS

I. No black spot are found at the tail end
Dorsal portion body is moderately humped
Scales on cheeks much larger than on body
Anal fin is a 100 - 110 fin ray ....................... *Notopterus notopterus*

II. A few black spots are present at the tail end
Dorsal portion body is relatively more humped
Scales on cheeks not much larger than on body
........................................................................... *Notopterus chitala*

## KEY TO THE ORDERS OF SUB-CLASS ACTINOPTERYGII

Abdomen non-keeled (except in genus Osteobrama) if keeled non-serrated.
Head scaleless body scaled or scaleless never covered by osseous plates.
Barbels are generally present ................................. CYPRINIFORMES

## Key to the Suborder of order CYPRINIFORMES

I. Body is scaly
Skin is covered with cycloid scales
Branchiostegals are 3-5 in numbers............................. CYPRINOIDEI

II. Body either nacked or covered with bony plates (scaleless)
Eyes are usually small
The barbels play important role in detecting the food
.................................................................................................. SILUROIDEI

## Key to the families of suborder CYPRINOIDEI

I. Body oblong or elongated
Branchiostegals three

    Scales are large and distinct
    Barbels if present 2-4 or absent
    Lateral line is usually present.........................................CYPRINIDAE

 II. Body is elongated, oblong, compressed or cylindrical
    Scales are minute and embedded in the skin
    Barbels are six or more
    Adipose fin is absent ..........................................COBITIDAE

## Key to the subfamilies of family CYPRINIDAE

  I. Abdomen is compressed and the ventral edge is cultural
    ..................................................................ABRAMIDINAE

 II. Abdomen is rounded
    Lateral line runs through the middle of the tail........CYPRININAE

III Lateral line runs close to ventral edge.......................RASBORINAE

IV. Abdomen strongly keeled from brest to vent.
    No barbels
    Scales do not extend to inter-orbital space
    ......................................................HYPOPHTHALMICHTHYINAE

## Key to the genera of subfamily ABRAMIDINAE

  I. Body is moderately elongated and compressed
    Abdominal edge is keeled but not serrated
    No knob at the symphysis of the lower jaw
    Lateral line curves abruptly downwards above the pectorals
    ..........................................................................*Salmophasia*

## Key to the species of genera Chela

  I. Ventral fin without an elongated outer ray
    L.I. – 34-37, A. 19-23 (2/17-21).........................*Salmophasia acinaces*

 II. Lateral line scale are from 86-110
    D. 9-11, P. 12-13, V. 9, A. 13-17............................*Salmophasia bacalia*

III. Lateral line curves gently downwards
    Lateral line scales are from 80-87
    D. 9, P. 13, V. 9, A. 18-19..............................*Salmophasia phulo*

IV. Bones of the pectorals girdle support the edge of the thorax
    Lateral line scales are from 140-160
    D. 9-10, P. 12-15, V. 8, A. 15-16..........................*Salmophasia sladoni*

**Key to the genera of subfamily *CYPRININAE***

 I. Upper lip and the barbels are absent
Lateral line is absent
Anal fin with seven rays
Abdomen is rounded and scales are small ........*Ambylpharyngodon*

 II. Eyes with free orbital margins
Upper lip and the barbels are absent
Lower lip is thick; scales are large
Lateral line is complete
Dorsal fin is with 17-19 fin rays.................................................... *Catla*

 III. Abdomen is rounded
Mouth is broad, transverse and sub-terminal
Upper lip is feebly fringed and is not continuous with the lower lip
One or two pairs of barbels are present
Lateral line is complete ............................................................. *Cirrhinus*

 IV. Abdomen is usually rounded
Body robust anteriorly, more or less compressed
Snout obtusely rounded
Lateral line straight with 36 scales......................................... *Cyprinus*

 V. Body elongated and sub-cylindrical
Upper lip and lower lips are continuous
A suctorial disc is present on the chin upper lip is fringed
One or two pairs of barbels are present
Lateral line is complete .................................................................*Garra*

 VI. Body is moderately elongated and abdomen is rounded
Lips are continuous at the angle of the jaws
Lateral line running along the middle of the side of the tail
One or two pairs of barbels are present................................... *Labeo*

 VII. Abdomen is rounded
No procumbent dorsal spine is found
Barbels are absent
Dorsal fin short and arises in between the bases of ventral and anal fins. It has an osseous serrated spine............................... *Osteobrama*

 VIII. Eyes without adipose lides
Lips are unfringed but the dorsal surface of the lower lip is no provided with tubercles
Both the lips are thin
Barbels may or may not be present....................................... *Puntius*

IX. Abdomen is rounded, head somewhat compressed
Upper lip is absent
Lateral line is complete
Scales are very minute
Barbels are absent
Anal fin is with 8 and the dorsal with 12 rays .......... *Thynnichthys*

X. Abdomen and snout is rounded
Upper jaw slightly longer than lower jaw
Barbels are absent ..................................................... *Ctenopharyngodon*

XI. Body is elongated and moderately compressed
Head is comparatively small
Lips are thick and continuous
The lower lips has transverse fold
Two pairs of barbels are present
Anal spine is not serrated .................................................. *Tor*

**Key to the species of genus *Ambylpharyngodon***

I. D. 9, L. . 55-60; five rows of scales between the lateral line and base of the ventral fin ........................ *Ambylpharyngodon microlepis*

II D. 9, L.I. 55-60; nine or ten rows scales between lateral line and base of ventral fin............................................*Ambylpharyngodon mola*

**Key to the species of genus *Catla***

I. D. 17-19, A. 8, L.I. 40-43; fins are blackish in colour

Scales have pink or coppery center except those of the ventral side, which are whitish.................................................................. *Catla catla*

**Key to the species of genus *Cirrhinus***

I. D. 10, A. 7, L.I. 48; two barbels are present
Dorsal rays 10................................................................*Cirrhinus fulungee*

II. D. 15-16, A. 8, L.I. 40-45; two barbels are present
Upper lip entire, dorsal rays 15-16........................*Cirrhinus mrigala*

III. D. 10-11, A. 8, L.I. 35-38; one pair of short rostral barbels
Upper lips indistinctly fringed or entire.....................*Cirrhinus reba*

**Key to the species of genus *Cyprinus***

I. Two pairs of barbels, upper jaw more or less projecting, but not protractile...........................................................................*Cyprinus carpio*

## Key to the species of genus *Garra*

I. D. 11, A. 7, L.I. 33-36; four barbels inter-orbital space convex
A black spot behind gill opening and generally a band along the
Sid ................................................................................................ *Garra lamta*

II. D. 10, A. 6, L.I. 35; four barbels, inter-orbital space flat
Five outer pectoral rays unbranched.................. *Garra lissorhynchus*

## Key to the species of genus *Labeo*

I. D. 11-13, L. tr. 6-7/7 barbels are 2, very short and maxillary
.................................................................................................... *Labeo boga*

II. D. 12, L.I. 60, L. tr. 12-14; one pair of maxillary barbels
.................................................................................................*Labeo boggut*

III. D. 16-18, L.I. 40-44, L. tr. 7.5/8, 4 barbels
Lips fringed with distinct inner fold
Slate or black colour........................................................*Labeo calbasu*

IV. D. 19-22, L.I. 44-47, L. tr. 9-10/8-9, 4 barbels
A thin cartilaginous layer on the inner side of both jaws
.................................................................................................*Labeo fimbriata*

V. D. 15-16, L.I. 40-42, L. tr. 6.5/9, 4 barbels
Lips thick fringed and with a distinct inner fold above and
below ........................................................................................*Labeo rohita*

## Key to the species of genus *Osteobrama*

I. D. 11, A. 14, L.I. 44, 4 barbels
Dorsal fins are 15 rows ............................................ *Osteobrama bakeri*

II. D. 11-12, A. 29-36, L.I. 55-70,
Osseous ray of the dorsal fin strong and serrated
.............................................................................. *Osteobrama cotio cotio*

## Key to the species of genus *Puntius*

I. D. 2-3/8, A. 2/5, L.I. 23-24, height of body 4.2-5 in total length
A dark lateral blotch sometimes present..............*Puntius amphibius*

II. D. 3/8, A. 2/5, L.I. 26-28, narrow sub-orbitals
A lateral blotch, and two bands on dorsal fin............*Puntius chola*

III. D. 3/9, L.I. 24-27, lateral line complete or incomplete
A black spot on side over anal fin ...................... *Puntius conchonius*

IV. D. 3/8, A. 2/5, L.I.21, without barbels
A black mark near the posterior end of lateral line
.................................................................*Puntius filamentosus*

V. D. 3/9, A. 3/5, L.I. 27-28, length of head 5 to 5.3 in total
Fins tinted orange and tipped with black.................*Puntius jerdoni*

VI. D. 3-4/8, A. 3/5, L.I. 32-34, height of body 3.5 – 3.7 in total
length
Lateral line complete, four barbels.............................*Puntius sarana*

VII. D. 3/9, A. 2/5, L.I. 25, dorsal spine weak
Without barbels...........................................................*Puntius sophore*

VIII. D. 3/8, L.I. 23-26, lateral line complete or incomplete

Two black spots, one at commencement of lateral line, another at
the side of the tail.................................................*Puntius ticto*

**Key to the species of genus** *Thynnichthys*

I. D. 12, A. 8, L.I.120, scales 17 to 19 rows between the lateral line
and the base of the ventral fin......................*Thynnichthys sandkhol*

**Key to the species of genus** *Ctenopharyngodon*

I. Dorsal fin without a spine
Lateral line continuous, slightly decurved
...........................................................*Ctenopharyngodon idella*

**Key to the species of genus** *Tor*

I. D. 12, A. 7-8, L.I. 25-27, four barbels
Height of body from 4.3 to 5.5 in the total length................*Tor tor*

**Key to the genera of subfamily RASBORINAE**

I. Sub-orbital ring of bones distinctly broad and prominent
Cleft of mouth often extending beyond anterior margin of eye
Abdominal edge is rounded
Barbels are four, two or none......................................................*Barilius*

II. Sub-orbital ring of bones not broad and prominent
Cleft of mouth not extending beyond anterior margin of eye
Abdominal edge is rounded
Barbels are two (rostral) or none...............................................*Rasbora*

## Key to the species of genus *Barilius*

I. D. 9, A. 13-14, L.I. 43-46, barbels 4
14 or 15 vertical bars ........................................................ *Barilius barila*

II. D. 9, A. 13-14, L.I. 39-42, and barbels are absent
9 vertical bands ............................................................. *Barilius barana*

III. D. 9, A. 9-10, L.I. 40-43, barbels 4
Short vertical bars each scale with a black spot in adults
.................................................................................... *Barilius bendelisis*

## Key to the species of genus *Rasbora*

I. D. 9 (2/7), A. 7 (2/5), L.I. 30-34, barbels are absent
Mostly a black lateral strip .................................... *Rasbora daniconius*

## Key to the genera of subfamily HYPOPHTHALMICHTHYINAE

Lateral line scales 110-115
Pelvic fins not very posterior ........................... *Hypophthalmichthyes*

## Key to the species of genus *Hypophthalmichthyes*

I. Anal fin with 14-17 rays
Dorsal fin inserted behind pelvic fins
.................................................................... *Hypophthalmichthyes molitrix*

## Key to the genera of family COBITIDAE

I. Body elongated and moderately compressed
A broad dark straight band runs along the lateral line
Caudal fin is entire or slightly emarginated
Six or eight barbels are present .......................... *Lepidocephalichthyes*

II. Body elongated, dorsal profile nearly horizontal
No erectile spine near the orbit
Six or eight barbels are present ......................................... *Nemacheilus*

## Key to the species of genus *Lepidocephalichthyes*

I. D.8, A. 7, L.I. 115 length of head ranges from 6.5 to 6.7
25 to 30 rows of scales between the base of the anal fin is forked
.................................................................. *Lepidocephalichthyes guntea*

**Key to the species of genus** *Nemachelius*

  I. D.11-12, A. 7, barbels 6, dorsal fin orange and rows of black spots
  Scales are indistinct ................................................ *Nemacheilus aeures*

  II. D.10, A. 7, dark bands, wider than the interspaces
  Caudal fin is forked ............................................ *Nemachelius beavani*

  III. D.11-14, A. 7, barbels 6 to 8
  Body irregularly blotched the caudal is slightly notched
  ..................................................................................... *Nemachelius botia*

## DIVISION – SILURI

**Key to the families of suborder SILUROIDEI**

  I. Head and body not ventrally flattened
  Paired fins not horizontal
  Anal fin is short and with less than 20 rays
  Adipose fin is always present and well developed ....... BAGRIDAE

  II. Body is elongated and the head is depressed
  Dorsal fin is spineless and very long
  Anal fin is short and with less than 45 rays
  Barbels are 4 pairs in number ........................................... CLARIIDAE

  III. Body is elongated and more or less laterally compressed
  Dorsal fin is spineless and very short
  Anal fin is very long with about 60-80 rays
  Barbels are 4 in number................................... HETEROPNEUSTIDAE

  IV. Body is elongated and compressed
  Dorsal fin has a spine
  Adipose fin is less developed or absent
  Anal fin moderate or long............................................. SCHILBEIDAE

  V. First dorsal is short and spineless
  Anal fin is very long
  Two pairs of barbels but no nasals..................................... SILURIDAE

**Key to the genera of family BAGRIDAE**

  I. Dorsal and pectoral spines are neither stout, firm nor hollow
  Ventrals have 6 fin rays
  Barbels are 8 in number................................................................ *Mystus*

  II. Dorsal and pectoral spines are stout, firm and hollow
  Ventrals have 8 fin rays
  Barbels are 8 in number................................................................. *Rita*

**Key to the species of genus *Mystus***

    I. A. 12-13, maxillary barbels extends beyond the caudal fin
A black spot, on the adipose dorsal fin is found .......... *Mystus aor*

    II. A. 9-10, maxillary barbels reaches the anal fin
Two light longitudinal bands and some times dark shoulder spots are found .......................................................................... *Mystus bleekeri*

    III A. 11-13, maxillary barbels reaches the caudal fin
The colour is silvery, often with a black spot at the base of dorsal spine ............................................................................. *Mystus cavasius*

    IV. A. 11-12, maxillary barbels reaches the fin
A black spot is found on the hind end of the base of the adipose fin ................................................................................. *Mystus seenghela*

    V. A. 11-13, maxillary barbels reaches the basal bone of the dorsal fin
Golden, with a black spot and about and five longitudinal dark bands .................................................................................... *Mystus tengra*

    VI. A. 9-12, four longitudinal bands present on each side of the body
Along with one median dorsal band ........................ *Mystus vittatus*

**Key to the species of genus *Rita***

    I. D. 7, A. 13-14, barbels 6, dorsal spine is very long serrated
Caudal fin is forked ................................................................. *Rita rita*

**Key to the genera of family CLARIIDAE**

    I. Head is depressed and the body is elongated
Dorsal fin is long and spineless
Adipose dorsal is absent,
Barbels are 8 in numbers ............................................................ *Clarias*

**Key to the species of genus *Clarias***

    I. D. 62-76, A. 48-58, barbels 8, vomerine teeth villiform
Body is brownish colour ............................................. *Clarias batrachus*

    II. D. 62-76, A. 48-58, barbels 8,

    Black longitudinal band on each side of the ventral side of the head ................................................................................. *Clarias gariepinus*

**Key to the genera of family HETEROPNEUSTIDAE**

I. Head is depressed and covered with thin skin
Mouth is transverse; dorsal fin is short or spineless
Caudal fin is rounded
Barbels are 8 in number..................................................*Hetropneustes*

**Key to the species of genus *Hetropneustes***

I. D. 6-7, A. 60-79, barbels 8
Body is dark leaden brown the young's are reddish
........................................................................... *Hetropneustes fossils*

**Key to the genera of family SCHILBEIDAE**

I. Cleft of mouth is oblique and extends below the middle of eye
Dorsal and pectoral fins are with spine
Anal fin possesses 29-51 rays
Barbels are 8 in number....................................................*Eutropichthys*

II. Rayed dorsal and adipose, both are present
Dorsal and pectoral fins are with spine
Lower jaw is pointed
Barbels are 4 in number...........................................................*Pangasius*

III. Body is elongated and compressed
Head covered with soft skin
Eyes with or without broad adipose lids
Barbels are 8 in numbers...............................................*Pseudeutropius*

**Key to the species of genus *Eutropichthys***

I. D. 8, A. 44-51, barbels 8, caudal fin is forked
Body is silvery grayish along the back............ *Eutropichthys vacha*

**Key to the species of genus *Pangasius***

I. D. 8, A. 31-34, barbels 8, pectorals spine is serrated
Body is silvery, darkest superiorly, with purple on sides
............................................................................*Pangasius pangasius*

**Key to the species of genus *Pseudeutropius***

I. D. 8, A. 43-42, barbels 8, maxillary barbels reach middle of pectoral
fin, the mandibular ones shorter than the head
.................................................................................*Pseudeutropius takree*

## Key to the genera of family SILURIDAE

I. Cleft of mouth neither too deep nor reaches to the anterior margin of eye
Barbels are 2-4 in numbers..............................................................*Ompak*

II. Cleft of mouth very deep extending beyond the posterior margin of eye
Barbels are 4 in numbers............................................................*Wallago*

## Key to the species of genus *Ompak*

I. D. 4, A. 60-75, barbels 4, and maxillary barbels reach ventral fin
Pectoral fin not so long as head
Spine smooth or serrated....................................*Ompak bimaculatus*

## Key to the species of genus *Wallago*

I. D. 5, A. 86-93, barbels 4, body is uniform silvery gray becoming lighter below.........................................................................*Wallago attu*

## Key to the genera of family SISORIDAE

I. Pectorals, dorsal and upper lobe of caudal produced into filaments
Body is nacked ........................................................................*Bagarius*

## Key to the species of genus *Bagarius*

I. D. 7(1/6), A. 13-15(3/10-12), barbels 8,
Caudal fin is deeply forked ....................................*Bagarius bagarius*

## KEY TO THE ORDERS OF SUBCLASS ACTINOPTERYGII

Body is elongated and nearly quadrangular in section
Fins are without spines
Lateral line is located very low, close to the belly
Branchiostegals are from 9-15 in number ..............BELONIFORMES

## Key to the suborder of orders BELONIFORMES

Both the jaws are produced to form a long beak
Teeth are long and needle like
The nasal is large and meet to each other along a suture
Scales are small......................................................SCOMBERESOCOIDEI

**Key to the family of suborder SCOMBERESOCOIDEI**

Body is elongated, sub-cylindrical
Both the jaws are much elongated with small and large needle like teeth
No detached fin lets and no gill rakers are found .... BELIONIDAE

**Key to the genera of family BELIONIDAE**

Eyes are lateral in position
Jaws are prolonged to form a beak
Dorsal fin originates opposite the anal fin origin .............*Xentodon*

**Key to the species of genus *Xentodon***

I. D. 15-18, A. 16-18, lateral line is not keeled
Scales are small and irregularly arranged ............. *Xentodon cancila*

**KEY TO THE ORDERS OF SUBCLASS ACTINOPTERYGII**

Head is depressed. Jaws are with teeth
Barbels and lateral line are absent
Fins are spineless
Pelvic fins are abdominal in position ..... CYPRINODONTIFORMES

**Key to the suborder of order CYPRINODONTIFORMES**

Anal opening is situated anterior to the anal fin
............................................................................CYPRINODONTOIDEI

**Key to the family of suborder CYPRINODONTOIDEI**

Caudal peduncle is longer than the head ....................POECILIDAE

**Key to the genera of family POECILIDAE**

I. Teeth conical and fixed.............................................................*Gambusia*

II. Teeth spatuliform and movable.................................................*Poecilia*

**Key to the species of genus *Gambusia***

I. D. 6-9, A. 8-10, L.I. 32
Sides of the body are irregularly dotted black .....*Gambusia affinis*

**Key to the species of genus *Lebistes***

I. This fish very similar to *Gambusia*
Caudal fin elongated (short and fan shaped).....*Poecilia reticulata*

### Key to the species of genus *Poecilia*

I. Brown spots on the side may have row of dusky black spots
............................................................................*Poecilia formosa*

## KEY TO THE ORDERS OF SUBCLASS ACTINOPTERYGII

Body is elongated with small to large scales
The scales may be cycloid or ctenoid
Two dorsal fins are present
They are costal and estuarine fishes, some entering freshwater
fishes............................................................................MUGILIFORMES

### Key to the suborders of order MUGILIFORMES

Mouth small and snout rounded
Lateral line absent or rudimentary
Teeth are not strongly socketted in the mouth...........MUGILOIDEI

### Key to the family of suborder MUGILOIDEI

Body is oblong or sub-cylindrical
Gill opening are wide, gills four
Scales are with a long perforation
Lateral line is absent............................................................MUGILIDAE

### Key to the genera of family MUGILIDAE

Eyes with or without an adipose lid
Teeth when present, minute
Branchiostegals four to six in number............................................*Mugil*

### Key to the species of genus *Mugil*

I. A. 3/8, L.I. 42-44, L.tr. 14, anterior and posterior eye lids are broad
Anal commences opposite the origin of second dorsal
............................................................................*Mugil cephalus*

II. A. 3/9, L.I. 48-52, L.tr. 15, extremity of maxilla visible
No long axillaries scale, 28 rows before dorsal fin
Anal commences opposite the origin of second dorsal fin
............................................................................*Rhinomugil corsula*

## KEY TO THE ORDERS OF SUBCLASS ACTINOPTERYGII

Body is elongated, sub-cylindrical anteriorly and compressed
posteriorly.

Fins are without spines
Dorsal and anal fins are long
Mouth is large and protractile
....................................CHANNIFORMES (OPHIOCEPHALIFORMES)

## Key to the family of order OPHIOCEPHALIFORMES

Branchiostegals 5 and pseudobranchiae absent
Head are snake like and having large shield like scales above
Fins are spineless; dorsal fin is single and long
Lateral line is abruptly curved or almost interrupted
.............................................................................. OPHIOCEPHALIDAE

## Key to the genera of family OPHIOCEPHALIDAE

Pyloric appendages are two in number
Ventral fins are present
Scales are large, moderate or small size
Anal fin is long but shorter than the dorsal ......................... *Channa*

## Key to the species of genus *Channa*

  I. D. 32-37, A. 21-23, L.I. 40-45, no bands are found on the body
     Dorsal, anal and caudal fins are tipped with orange
     ......................................................................................*Channa gaucha*

 II. D. 45-55, A. 28-36, L.I. 60-70, and the bands like that of C. striatus
     are not found but a large black ocellus is present at the upper part
     of the base of the caudal fin ....................................*Channa marulius*

III. D. 29-32, A. 21-23, L.I. 35-40, greenish gray becoming yellow on
     the ventral side
     Several short bands or patches run from the back and pass down
     the abdomen, fins are spotted ...............................*Channa punctatus*

 IV. D. 37-45, A. 23-26, L.I. 50-60, bands of gray and black descend from
     the lateral sides to the abdomen
     There is no ocellus on the tail....................................*Channa striatus*

## KEY TO THE ORDERS OF SUBCLASS ACTINOPTERYGII

Body is oblong or elevated but very rarely elongated
Usually two dorsal fins are present
Scales are ctenoid or anterior cycloid ......................... PERCIFORMES

## Key to the suborder of order PERCIFORMES

I.  Accessory branchial organs are present
Body is neither cutlass like nor the tail is tapering
Anal fin has more than 3 spines (8-20) ................. ANABANTOIDEI

II. Accessory branchial organs are present
Eyes are dorsal in position
Pelvic fins have united to form a sucking disc
There is a gap between the spiny and soft dorsal fin
............................................................................................... GOBIOIDEI

III. Head is not depressed and is without spines and bony ridges
There is no gap between the spiny and soft dorsal fin
Eyes lateral in position ......................................................... PERCOIDEI

## Key to the family of suborder ANABANTOIDEI

I.  Body is compressed, oblong or elevated
Mouth is relatively large,
Eyes are lateral in position
Lateral line is either interrupted or absent
Scales are of ctenoid type ........................................... ANABANTIDAE

## Key to the genera of family ANABANTIDAE

Opercle and preorbitle serrated
Palatine teeth are present
An accessory branchial organ is present
Ventral fin is normal ......................................................... *Anabus*

## Key to the species of genus *Anabus*

I.  D. 17-18/8-10, A. 9-10/9-11, L.I. 18-32,
There are four wide bands on the body. A black blotch is present
on each side of the caudal peduncle ................... *Anabus testudineus*

## Key to the family of suborder GOBIOIDEI

Body is generally elongated
Mouth is usually larger with small canine teeth
Two dorsal fins are present
Lateral line may or may not be present ........................... GOBIIDAE

## Key to the genera of family GOBIIDAE

Head is depressed, pointed and scaled behind the eyes
Two dorsal fin are present first 6 rays and second 1/6-10 rays
Pelvic fins are united to form a weak-sucking disc
Base of the pectoral fin is scaled ..................................... *Glossogobius*

## Key to the species of genus *Glassogobius*

I. A. 9-10, L.l. 30-35, L.tr. 8-12, dorsal fin are two in number first
dorsal with 6 weak spines
Caudal fin is oblong ............................................. *Glossogobius giuris*

## Key to the family of suborder PERCOIDEI

I. Dorsal fin is with 7 and anal with 3 spines
No scaly process is found in the axil of ventral
Body is glass like
Scales are of cycloid type ................................................ AMBASSIDAE

II. Dorsal fin is with 12 or more spines
Maxilla extends beyond eye
Both the dorsal fin is continuous ..................................... NANDIDAE

III. Anal fin is with 3 or more spines
Rays are not developed covered by the skin
Abdomen is not sharply edged
Body is oblong and laterally compressed ...................... CICHLIDAE

## Key to the genera of family AMBASSIDAE

Body is compressed. Branchiostegals are 6 in number
Pseudobranchiae are present
Dorsal fins are two in number, first 7 spines
Lateral line may be complete, or incomplete broken or even
absent ............................................................................... *Chanda*

## Key to the species of genus *Chanda*

I. D. 1+7/1/13-17, A. 3/14-18, an oblong vertical patch of black colour
is present on the shoulder
Anal fin has spots at the base of the spine ................. *Chanda nama*

II. D. 1+7/1/13-15, A. 3/14-16, lower jaw is without large and curved
caniniform teeth
A dark shoulder spot is present ........................... *Parambassis ranga*

## Key to the genera of family NANDIDAE

Body is oblong and compressed. Head is usually large
Dorsal fin is single
Anal fin is with 3 spines
Lateral line is complete or incomplete ......................................... *Badis*

## Key to the species of genus *Badis*

I. D. 16-18/7-10, A. 9-11 (3/6-8), dorsal fin is with more than 14 spines
Preopercle is not serrated...................................................*Badis badis*

## Key to the genera of family CICHLIDAE

Body short, more or less elongate
Abdomen is rounded
Scales cycloid (rarely indistinctly ctenoid)
Anal fin with 3 or 4 spines ............................................... *Oreochromis*

## Key to the species of genus *Tiliapia*

I. Dorsal fin with 15 or 16 spines and 10 or 11 rays
Spinous portion longer than soft part ..... *Oreochromis mossambica*

## KEY TO THE ORDERS OF SUBCLASS ACTINOPTERYGII

Body is elongated and eel like
Dorsal, caudal and anal fins are confluent
Nasals are very long and meet in the mid line
Branchial openings are small and ventral in position
......................................................................... MASTACEMBELIFORMES

## Key to the family of order MASTACEMBELIFORMES

They are freshwater eel like spiny fishes
Pseudobranchiae are absent
Dorsal fin is long
Pectoral fins are present but the pelvic fins are absent
Anal fins has 3 spines .........................................MASTACEMBELIDAE

## Key to the genera of family MASTACEMBELIDAE

The snout has no transversely striated on its ventral side
A conspicuous preorbital spine is present
Black ocelli are not found along the base of the dorsal fin
.............................................................................*Mastacembelus*

**Key to the species of genus *Mastacembelus***

I. D-32-39/74-90, A. 78-91 Anal and dorsal fins are confluent with the caudal
Body colour is greenish, marbled, spotted, with or without undulating lines...................................................................***Mastacembelus armatus***

II. D. 24-26/30-42, A. 3/31-46, Anal and dorsal fins are not confluent with the caudal
Sides are spotted with yellowish white colour
Vertical fins are with fine black spots.........***Macrognathus pancalus***

# 2

# MORPHOLOGICAL VARIATION

Teleost fish show great variation of body form, fins, scales and other characters of body. This variation is great taxonomic value, particularly at the lower taxonomic units.

## 2.1 BODY FORM

(i) Asymmetrical body: Some fish lakes flat are exceptional in having a bilaterally asymmetrical body.

(ii) Symmetrical Body: Majority of fish has a bilaterally symmetrical body form.

(iii) Compressed form: The profile of the body is laterally compressed, *e.g.,* Butterfish.

(iv) Depressed form: The body of fish is flat dorsoventrally, *e.g.,* Cat fish.

(v) Ribbon Like form: The body of fish is elongated but compressed laterally, *e.g.,* Ribbonfish.

(vi) Globiform: The profile of the body is globe like, *i.e.,* round in cross-section.

vii) Filiform: The body is thread like, *i.e.,* highly off enuated profile.

viii) Fusiform: The fish is streamlined or boat shaped the contour gently sweeps back from a maximum girth in cross section at about $\frac{1}{3}$ of the length from the anterior end, *e.g.,* Tuna.

ix) Serpentine form: The body is snake like *i.e.* cylindrica, elongate and subcylindrical in cross section, *e.g.,* Eel.

Eel-like, greatly elongated, attenuated

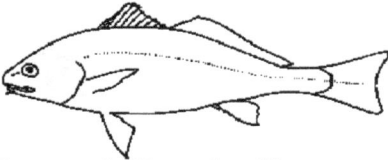

Elongate, fusiform, basslike

Ovate, truncated

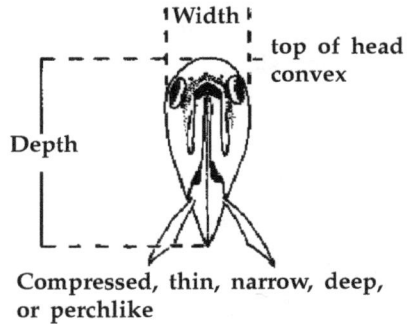

Width

top of head
convex

Depth

Compressed, thin, narrow, deep,
or perchlike

Body depressed, flattened

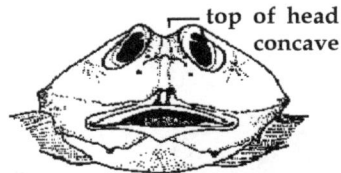

top of head
concave

Body subcircular, hemispherical

Figure 2.1: Some Body Form of Fishes

## 2.2 FINS

The fin variation in structure with regard to the shape, position, number of rays or spines. The basis of variation is the varying extends which a fin plays role in locomotion and maneuvering. Fins are the chief organs of locomotion in fishes and are of two kinds.

(i) Median or unpaired (ii) paired

The median fins include [dorsal, caudal and anal] the dorsal on the back, on anal on the ventral side behind the vent, and caudal at the end of the tail.

The paired fins include (pectoral and pelvic fin) the pectoral and pelvis fins corresponding to the fore and hind limbs of terrestrial vertebrates.

A fin may be simple, provided with rays or spines or both. An adipose fin is fleshy and devoid of rays or spines. Some variations obtaining in major fins are given below.

(A) **Dorsal Fin**: The number may be one, two or even three (cod.) The dorsal fin may be rudimentary or altogether absent or greatly elongated. An elongated dorsal fin may occur as single fin (*Clarias*) or broken into a number of isolated fin lets.

B) **Anal Fin**: It may be normal sized or greatly enlarged or completely absent. In *Notopterus* it may be enlarged into an elongate fin for propulsion. The anal fin may be modified into an intromittant organ called gonopodium.

C) **Caudal Fin**: The caudal fin when present display a great variety of shapes some of the important variations are shown in fig.

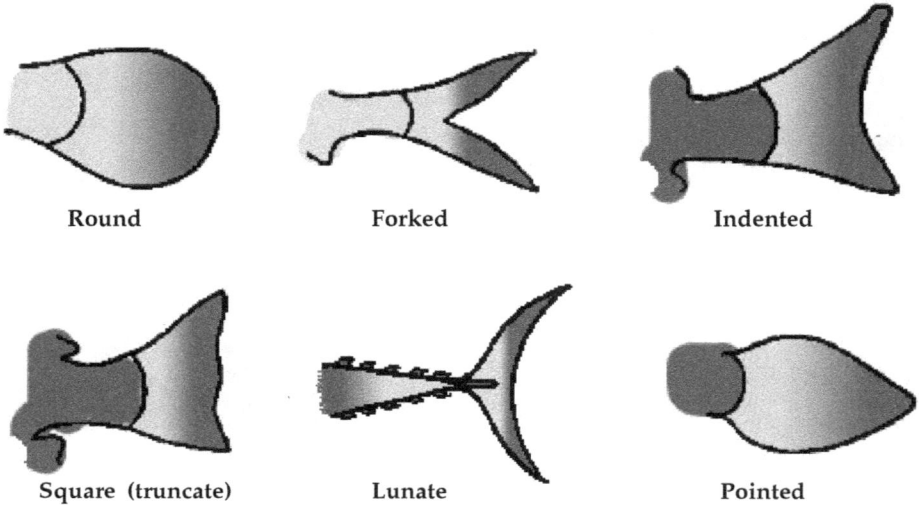

|  |  |  |
|---|---|---|
| Round | Forked | Indented |
| Square (truncate) | Lunate | Pointed |

**Figure 2.2: Variation in Shape of Caudal Fin**

(D) **Pectorals Fin**: The fin usually found on each side of the body behind the gill opening; this pair of fins is found on the lower parts of the body in primitive forms of fish; corresponding to the forelimbs of higher vertebrates; united to form the disc in most rays. The pectorals fins are used for steering to change direction and as a break to slow down or stop the movement.

(E) **Pelvic or Ventral fin:** The paired fin which is located posterior, ventral or anterior to the pectoral fins (abdominal, thoracic or jugular in position). It functions to steer, brake and propel the fish and acts as a keel. In the pelvic fin ray count usually all the rays are counted except a small ray preceding the first ray and usually bound so closely to it so as to require dissection to be seen.

## 2.3. MOUTH, SNOUT, JAWS AND TEETH

Some important variations are noted below:

The mouth may be terminal, superior or inferior snout may be pointed, rounded, tubular and overhanging the mouth jaws may be normal or projecting. When projecting, either both jaws are projecting or only one of the jaws (upper or lower). Teeth on one jaw may be cardiform (short and pointed), Villiform (large pointed) canine like (piercing, doog tooth or fang like), incisor (sharp cutting), comb like or molariform (crushing and grinding). Teeth may be found on sites other than jaws, such as vomerine teeth (on vomer bone), pharyngeal teeth (on gill arches) and tongue teeth (on the tongue).

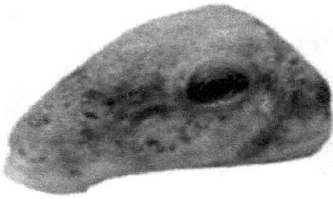

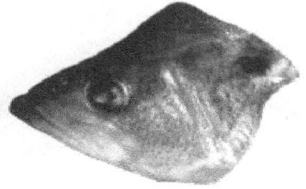

Inferior mouth          Terminal mouth          Superior mouth

**Lower Jaw Projecting beyond Upper Jaw**

**Upper Jaw is Prolonged into a swordlike beak**

**Snout *Tubular* with Jaws at tip**

**Jaws (and Lips) are *Terminal, i.e.,* at end of body**

**Snout *Overhanging* or *Projecting* beyond Mouth, the Mouth is thus *Inferior***

**The Upper Jaw is *Extended* and the Lower Lip is *Inferior* or *Included***

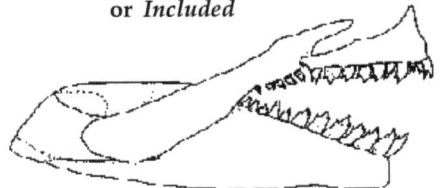

**Conical teeth**

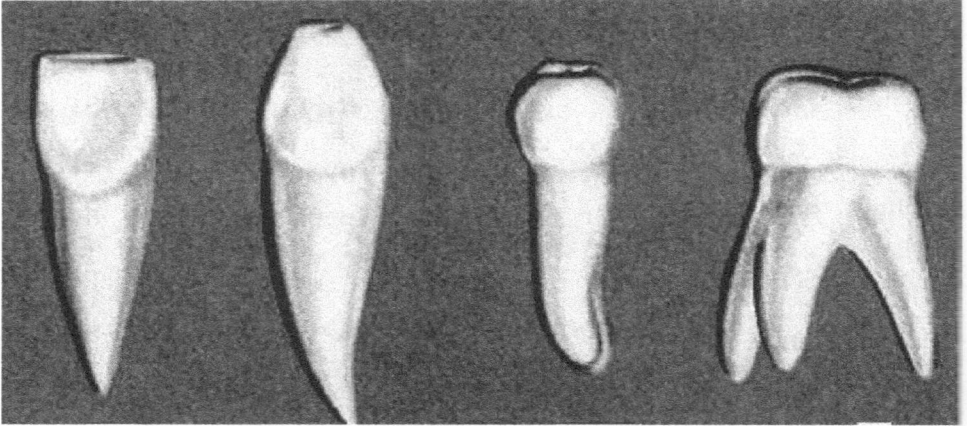

Incisor, Canine, Premolar and Molar Teeth

Figure 2.3: Mouth, Snout, Jaws and Teeth

## 2.4. BODY COLOURATION AND SCALES

Colouration in fishes is due to the presence of two types of special cells called the chromatophores and iridocytes. The chromatophores are branched connective tissue calls situated in the dermis either above or below the scales. These cells contain various kinds of pigment granules, which may be Carotenoids (Yellow, Red), Melanins (Black), Flavines (Yellow), Purines (White or Silvery), Pterine (White), Porphyrins and bile (Red, Yellow, Green, Blue, and Brown), Inigoid (Blue, Red, Green), Chromolipoid (Yellow to Brown). Depending upon the colour of the pigments the chromatophores are designated as:

- Eryrhrophores (Red or Orange)
- Xanthophores (Yellow)
- Melanophores (Black)
- Leucophores (White)

The iridocytes contain a crystalline substance guanine, which is opaque, whitish or silvery (mirror hue).

A large number of Teleostean fishes are brightly and brilliantly coloured others are of more uniform and sober shade Colouration is mainly due to skin pigments but the background colour may also be due to underlying tissue and body fluids.

**Scales:** Scales are integumentary derivatives. Their types, distribution, modification or absence as well as scale counts have diagnostic value in taxonomy.

Scalation presents a great range from an imbricating pattern (carps) to mosaic pattern. However, scales are completely absent in most calfish. Four types of scales are presents in fishes as follows.

(i) Tooth like (placoid scales)

(ii) Diamond shaped (rhombic scales)

(iii) Disc like with caudal exposed margin smooth (cycloid scales)

(iv) Disc like with caudal exposed margin toothed (ctenoid scales)

Scales may undergo modification, often beyond recognition, some glaring examples of modification include fin spines and rays, and teeth on snout of saw fish, scutes on the belly of clupeids. Scales may develop as thick plates covering the body, forming dermal armour, on encasement of body in a rigid case or a semi rigid case.

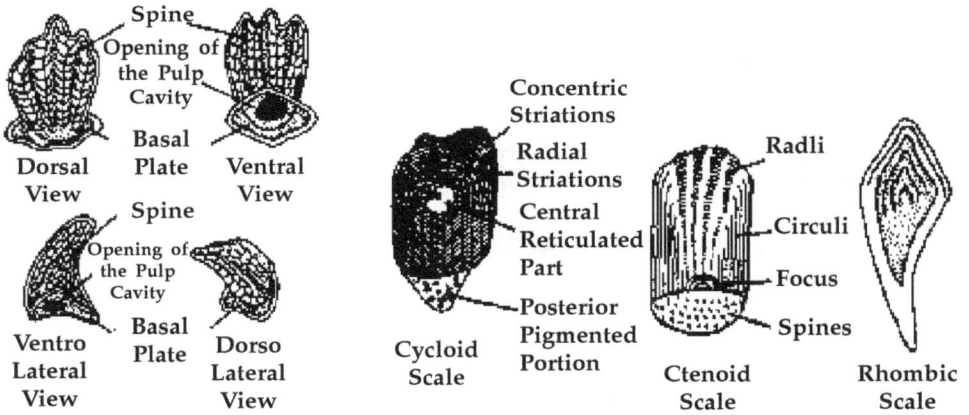

Spine
Opening of
the Pulp
Cavity
Basal
Plate

Dorsal          Ventral
View             View

Dorsal Plate Ventral

Spine
Opening of
the Pulp
Cavity
Basal
Plate

Ventro          Dorso
Lateral          Lateral
View             View

Concentric
Striations

Radial
Striations

Central
Reticulated
Part

Posterior
Pigmented
Portion

Cycloid
Scale

Radli
Circuli
Focus
Spines

Ctenoid
Scale

Rhombic
Scale

Figure 2.4: Types of Scales

# 3

# MEASUREMENTS AND COUNTS USED FOR THE DIAGNOSES

In making an account of fish identification, certain terms and abbreviation are used to describe the various parts and measurements of the body of the fish and meristic characters are those characteristics, which can be counted such as scales fins and fin rays etc. The following terms are of common use, which may be explained as under.

## 3.1 MORPHOMETRIC CHARACTERS

**Total length (T.L.):** The entire length of the body of a fish is known as its total length. From the tip of snout with the mouth closed to the tip of the longest ray in the caudal fin

**Standard length (S.L.):** It differs from that of the total length in respect of that it is measured from the tip of snout up to the point of origin of the caudal fin, *i.e.*, up to the end of the last vertebra of the vertebral column.

**Fork length (F.L.):** It is the distance from the tip of snout with the mouth closed to the anterior most point of the caudal posterior margin.

**Length of head (H.L.):** It is a distance, measured from the tip of the snout of the posterior-most point of operculum.

**Length of head excluding the snout (SHL):** It begins from the anterior margin of the orbit and extends up to the posterior most bony extremity of the operculum.

**Length of Snout (SNL):** It is the distance from the tip of the snout up to the anterior margin of the orbit is said to be the length of the snout.

**Postorbital length of the head (POL):** It is the distance in a straight line between the posterior margin of the orbit and the posterior most edge of the opercular bone.

**Length of caudal peduncle (CPL):** It is the distance, which is measured from the posterior base of the anal fin to the origin of the caudal fin.

**Height of body (HB):** It is measured the vertical distance of the body at its deepest part.

**Height of caudal peduncle (CPH):** It is measured vertically through the body, at its narrowest part.

**Head depth:** the vertical distance of the head at its deepest part, measured from the middle at the occiput (the posterior part of the head) vertically downwards to the external contour of the head.

**Head width:** The distance from side-to-side at the widest part of the head when the opercules are forced in a reasonably normal position.

**Interorbital width:** The distance, along the dorsal surface of the head lying in between the two orbits is known as interorbitals space.

**Diameter of the eye (ED):** It is measured from one margin of the orbit to the other.

**Predorsal length:** It is the distance between the anterior most ends of the body and the front end of the dorsal fin base.

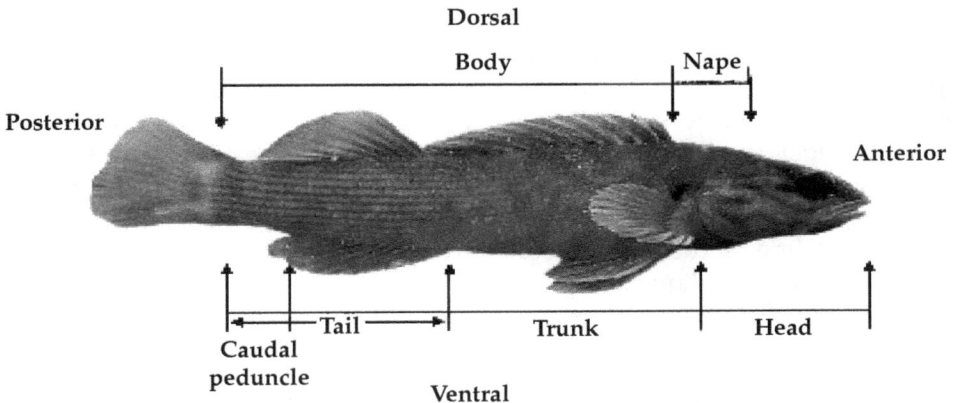

Figure 3.1: The topography showing a scaly fish of its different fins, scales and measurements.

**Length of upper jaw:** From the tip of snout to the posterior edge of the axillary's bone.

**Length of lower jaw:** It is the greatest length of lower jaw.

**Measurement of fins:** The length of pectoral, pelvic and the caudal fins is measured along their longest fin rays. The height of dorsal and anal fins is measured along their longest rays or spines but the length is measured along their total stretch, beginning from the first fin ray of the spine up to the last fin ray.

**Profile of the body:** It explains the outline of the body of fish along its dorsal as well as the vertical surfaces.

**Barbels:** From the anterior base of the barbels to its tip when it is straightened. Barbels are slender and elongated structures which are found around the mouth of many fishes. The number of barbels may very from 1-4 pairs. According to their position, they are named as nasal, rostral, maxillary and mandibular.

**Lateral Line (L.I.):** The lateral line is a longitudinal row of perforations special sense organs, on the lateral sides of many fishes; it may be complete, incomplete or interrupted.

**Branchiostegeal rays:** These are slender bony rods, which are found on the inner surface of the operculum.

**Scales:** According to the mode of their origin, these are two types of scales. (i) Those, which are formed due to the secretary activity of both epidermis and dermis, as the placoid scales if elasmobranch and (ii) Non-placoid scales that are derived from the dermis only as the scales of teleosts. The number of scales along the longitudinal and transverse rows is an important tool in the identification of scaly fishes. If the lateral line is present the scales are counted along its length and their number is written after the abbreviation L.I. In case the lateral line is absent, the scales are counted along the row where the lateral line might has been located and there number is written after the abbreviation L.r. the transverse rows of scales are counted from the anterior base of the dorsal fin in the ventral line and their number mentioned after the abbreviation L.tr. In this case, if a lateral line is present the scales above and below the lateral line are separated by an oblique (/) stroke predorsal scales are counted from the extremity to the origin of the dorsal fin.

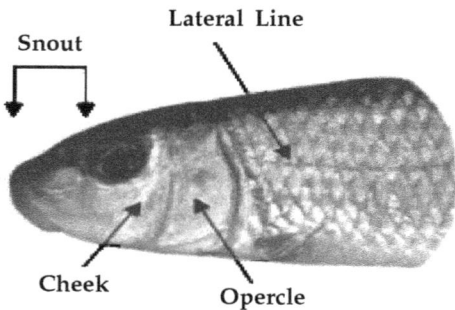

Figure 3.2: Lateral view of head        Figure 3.3: Front View of the head

## 3.2. MERISTIC CHARACTERISTICS

The standard abbreviations are used as follows:

**Abbreviations**

|       |                                      |
|-------|--------------------------------------|
| A     | Anal fin                             |
| B     | Branchiostegal rays                  |
| C     | Caudal fin                           |
| D     | Dorsal fin                           |
| O     | Adipose fin                          |
| P     | Pectoral fin                         |
| V     | Ventral or Pelvic Fin                |
| L.I.  | Lateral line of Perforated scales    |
| L.r.  | Lateral row of unperforated scales   |
| L.tr. | Lateral transverse rows of scales    |
| FMA   | Freshwater of Marathwada region      |

Examples of some meristic counts are presented below:

## 3.3. FINS AND SCALES FORMULA

The above abbreviations mentioned are used in the construction of fins and scales formulae of fishes. These formulae provide scientific information to confirm the actual identity of a particular fish. Every abbreviation carries a number or number after it which denotes the number of fin rays, scales etc. Number of fin supports; both types of fin supports; spines, rays for each fin were counted.

**Scales count:** Scales in the lateral line from the first scale touching the shoulder girdle and terminating at the hypuralpoint. Which a hyphen (-) appear in between the two numerical figures it indicates the range of variation. An oblique (/) stroke and separates two types of range in a single fin such as spiny and unbranched rays from the branched once while a vertical ( | ) stroke separates different fins from each other, such as rayed dorsal from the adipose dorsal or the spiny fin from the non spiny one. Thus the fins and scales formula, "D.1/7/0, P.1/8-9, V.6, A. 4/7-9, C. 16 Barbels 4 pairs." Siluroid fish explains as under figure 3.4.

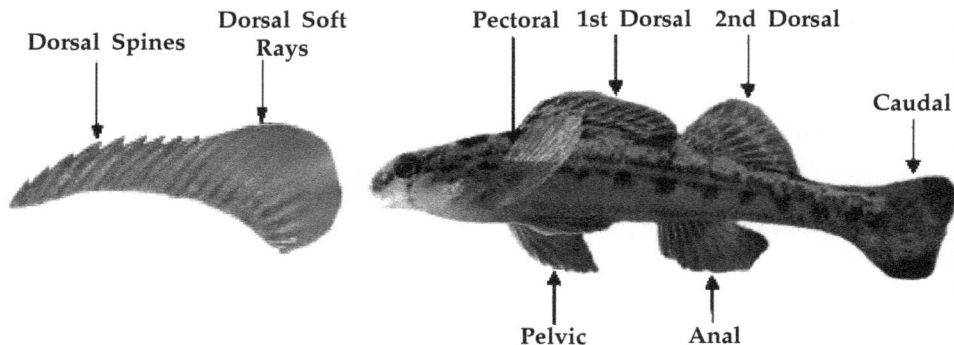

Figure 3.4. A Siluroid fish, showing the topography of different fins and other structures.

**D (Dorsal Fin) 1/7 | 0:** The dorsal fins are two in number of same fishes 1/7 is separated from second by a vertical stroke .The first dorsal fin possesses as a whole, 8 fin rays out of which the first fin rays are spinous (unbranched) and the 7 fin rays are soft (branched) and hence the above figure pertaining to the same fin are separated by an oblique stroke.

**P (Pectoral Fin) 1/8:** Whole fin rays 9 out of which the 1st one is spinous and the rest 8 are soft ones.

**V (Ventral or pelvic fin) 6:** Possesses 6 fin rays of one type only.

**A (Anal fin) 4/7-9:** Consist of in all 11 to 13-fin rays out of which four are spinous and the rest 7 or 9 are soft ones.

**C (Caudal Fin) 17:** Has 17 fin rays of one type only.

**Barbels 4 pairs:** The fish possesses 4 pairs of barbels and some fishes of two pairs.

In the above example, the abbreviations L.I., L.r. and L.tr. are not met with as in Siluroid fishes the scales are absent. This aspect along with the others can be explained with the help of a formula pertaining to a scaly fish. Thus the following formula:

"D.10 (2/8), P.15-16, V.9, A. 7(2/5), C.19, L.I. 46-48, L.tr.8/9." These formula explains that figure 3.2.

**D (Dorsal fin) 10 (2/8):** The dorsal fin possesses as a whole, 10 fin rays out of which 2 is spinous and the rest 8 fin are soft. Hence the presence of an oblique stroke in between the two numerical figures

**P (Pectoral fin) 15-16:** Consist of 15-16 fin rays of one type only.

**V (Ventral or pelvic fin) 9:** Possesses 9 fin rays of one type only.

**A (Anal Fin) 7(2/5):** Has total number of fin rays 7 but out of this number two are spinous and rest five (5) are soft.

**C (Caudal Fin) 19:** Consist of 19 fin rays of one type only.

**L.I. (Lateral line of perforated scales):** The number of perforated scales are 46-48 along the length of the lateral line varies from 46 to 48.

# 4

# SYSTEMATIC SECTION

| Class | - Teleostomi |
|-------|--------------|
| Order | - Clupeiformes |
| Sub-order | - Notopteroidei |
| Family | - Notopteridae |
| Genus | - *Notopterus* |

Synopsis of the species of the genus *Notopterus* Lacepede dealt within this text.

**1. *Notopterus chitala* (Ham.) 1822**

B.8-9; D.9-10 (1-2/7-9), P.16; V.6; A. 110-125; C.12-14; L.I. 160-180.

| | |
|---|---|
| Popular names | : Feather back, Knife fish. |
| Local names | : Pholi, Patola, Chapple-Mache Chambhari. |
| Environment | : Demersal fresh water pH range 6.0 to 8.0 dh range 5.0 to 19.0 |
| Climate | : Tropical 24-25. |
| Dangerous | : Harmless |

**Characters**

1. Length of Head ranges from 4.5 to 5 and height of body 3.5 to 4 in the total length.

2. Mouth large, Maxilla extends beyond hind edge of hide edge of eye.

3. Dorsal fin originates nearer the tail than the snout.

4. Anal fin is long and confluent with the caudal latter somewhat pointed Ventrals are minute.

5. Body is coppery brown or grayish, Silvery on sides.

6. Lateral line is complete.

Remark                : It averages 25 cm in length but may reach up to meter.

Importance            : Fisheries: minor commercial; aquaculture: commercial; Game Fish: yes; Aquarium: commercial.

Maximum size          : 122 cm SL.

**2. *Notopterus notopterus* (Pallas) 1769**

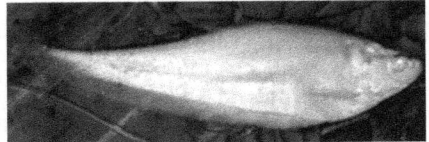

B. 8; D. 7-8 (1-2/6-7); P. 17; V. 5-6; A. 100-110; C. 19; L. I.225

Popular names         : Feather back or knife fish.

Local names           : Pholi, Patola, Chapple machi, Chambhari.

Environment           : Demersal; freshwater; Brackish; pH range 6.0 to 6.5 dH range 3.0 to 8.0

Climate tropical      : 24-28°C; 3.5°N-10°S.

Dangerous             : Harmless

**Characters**

1. Length of head ranges from 4.7 to 5.5 and the height of body from 3.5 to 4 in the total length.

2. Mouth moderate; Maxilla extends up to mid orbit.

3. Body is deep, moderately elongated and compressed.

4. Dorsal fin is small and arises nearly midway between the snout and end of the caudal fin. Ventral are very minute.

5. Anal originates just behind the ventral and becomes confluent with caudal fin.

6. Caudal fin is pointed.

7. Lateral line is complete.

8. Body is silvery white, with fine gray spot, which are dark along the narrow back.

Remark : This fish is relished both in fresh and dried condition. Owing to its carnivorous nature.

Importance : Fisheries: commercial; aquaculture; commercial. Aquarium: commercial.

Maximum size : 60.0 cm SL.

**ORDER-CYPERINIFORMES** (This order is divided into two orders)

(1) Division - Cyprini.

(2) Division - Siluri.

Division : Cyprini

Suborder : Cyprinoidei

Family : Cyprinidae

Subfamily : Abramidinae

Genus : *Salmophasia*

Synopsis of the species of the genera *Salmophasia* dealt within this text.

**3. *Salmophasia acinaces* (Valenciennes, 1844)**

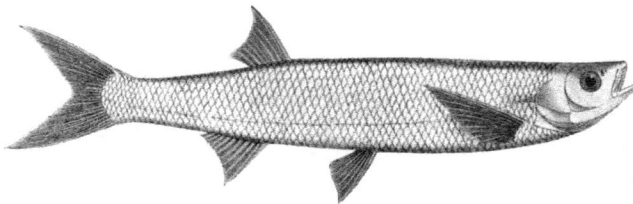

B. 3; D. 9-10 (2-7/3-8); P. 15; V. 8; A. 17-19 (3/14-16); C. 19; L.I. 43-45; L.tr. $6_{1/2}$-7/3.

| | |
|---|---|
| Popular names | : Minnows or carps |
| Local names | : Chelliah, Chalti |
| Environment | : Benthopelagic, freshwater |
| Climate | : Tropical |
| Dangerous | : Harmless |

**Characters**

1. Length of head ranges from 5½ to 5⅔ and the height of body from 5 to 5½ in the total length.

2. Body is elongated and compressed.

3. Cleft of mouth extending to below the anterior third of the orbit.

4. Dorsal fin situated in the posterior third of the body.

5. Pectoral reaches the ventral.

6. Caudal fin is deeply lobed.

7. Lateral line descends gently for first twelve scales.

8. Colour of the body is silvery. Caudal dark edged.

| | |
|---|---|
| Remark | : It is a fast swimmer and found in clear water. |
| Importance | : Fisheries of no interest. |
| Maximum size | : 15.0 cm TL. |

### 4. *Salmophasia bacaila* (Hamilton, 1822)

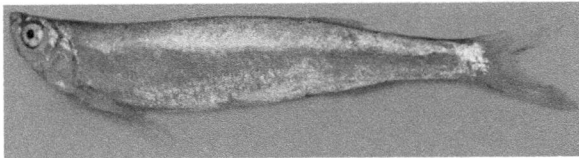

B. 3; D. 9-11 (2/7-9) 0; P. 12-13; V. 9; A. 13-17 (2/11-15); C. 19; L.I. 86-110; L.tr. 17-7-19\6-10-11\6-10.

| | |
|---|---|
| Popular names | : Minnows or carps |
| Local names | : Chelliah |
| Environment | : benthopleagic, freshwater, brackish |
| Climate | : Tropical |
| Dangerous | : Harmless |

**Characters**

1. Length of head range from 5.2 to 6 and the height of body from 5.5 to 6 in the total length.

2. Body is elongated and compressed.

3. Cleft of mouth touches the first fourth of the eye.

4. Mouth is directed upwards.

5. First anal ray is middle of the dorsal fin.

6. Pectorals are elongated. Caudal fin is forked.

7. Lateral line is complete with a downward curve at the pectoral fin.

8. Scales are small and thin.

9. Colour of the body is silvery white.

Remark          : It provide good dish for poor meal.

Important       : Fisheries of no interest.

Maximum size    : 18.0 cm TL.

**5. *Salmophasia phulo* (Hamilton, 1822)**

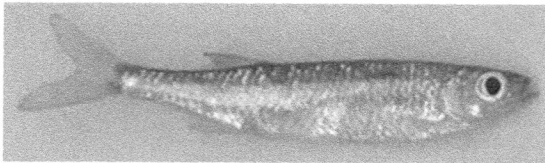

B. 3; D. 9 (2-7); P. 13; V. 9; A. 18-19 (2-3/16); C. 19; L.I. 80-87; L.tr. 12/15/6.

Popular names   : Silverfish, minnows or carps

Local names     : Dannahree, Chela phulo

Emvironment     : benthopelagic, freshwater

Dangerous       : Harmless

**Characters**

1. Length of head is 5.5 and the height of body range from 4.5 to 5 in the total length.

2. Body is elongated and compressed.

3. The maxilla reaches to below the front margin of the orbit.

4. The dorsal fin arises apposite the origin of the anal fin.

5. The caudal is deeply forked and its lower lobe longer.

6. Lateral line curves gently downwards.

7. Colour of the body is silvery with a bright lateral bond on both sides.

Remark                :  It grows up to 13 cm in length.

Importance            :  Fisheries of no interest.

### 6. *Salmophasia sladoni* (Day, 1870)

B. 3; D. 10 (2-8); P. 11; V. 8; A. 20-21 (2/18-19); C. 21; L.I. 65-68; L.tr. 10/8.

Popular names        :  Minnows or carps

Local name           :  Chalte

Environment          :  Benthopelagic, freshwater

Climate              :  Tropical

Dangerous            :  Harmless

### Characters

1. Length of head 6 to 6¼, and height of the body ranges from 5 to 5½ in the total length.

2. Body is elongated and compressed.

3. Maxilla reaches of the posterior extremity to the anterior third of the orbit.

4. Dorsal fin commences opposite the anal fin.

5. Pectorals as long as the head but does not reach the ventral.

6. Body of the colour is silvery, caudal black edged.

Remarks            : It grows up to 12 cm total length.

Importance       : Fisheries of no interest.

Subfamily         : CYPRININAE

Genus               : Ambylpharynygodon (Bleeker) 1860

Synopsis of the species of the genus Ambylpharyngodon Bleeker dealt within this text.

### 7. *Ambylpharyngodon microlepis* (Bleeker) 1853

B. 3; D. 9 (2-7); P. 14; V. 9; A. 7 (2-5); C. 19; L. I55-60; L.tr. 11/8.

Popular names    : Minnows or carps

Local names       : Dhawali

Environment      : Benthopelagic, freshwater

Climate            : Tropical

Dangerous        : Harmless

## Characters

1. Length of head range from 4.7 to 5 and the height of body from 4.5 to 5 in the total length.

2. Abdomen is rounded. Head is slightly compressed.

3. Dorsal fin originates slightly behind the origin of the ventral fins

4. Caudal fin is forked with the lower lobe slightly longer.

5. Lateral line is complete.

6. Barbels are absent. Scales are of small size.

7. Colour of the body is silvery band is present on each lateral side of the body.

Remarks            : It grows up to a length of 10 cm.

Importance       : Fisheries of no interest.

## 8. *Ambylpharyngodon mola* (Ham.) 1922

B. 3; D. 9 (2-7); P. 15; V. 9; A. 7 (2-5); C. 19; L.I. 65-75; L.tr. 11/12.

| | | |
|---|---|---|
| Popular names | : | Minnows or carps |
| Local names | : | Mola carplet |
| Environment | : | Benthopelagic, freshwater |
| Climate | : | Tropical |
| Dangerous | : | Harmless |

### Characters

1. Length of head is 5 and the height of the body ranges from 4 to 4.5 in the total length.
2. Abdomen is rounded. Head is slightly compressed.
3. Mouth is anterior, the lower jaw being slightly more prominent and elongated.
4. Dorsal fin profile is more convex than that of the abdomen.
5. Caudal fin is forked.
6. Lateral line is interrupted.
7. Colour of the body is silvery lateral band is present on either side of the body and small black dots on the upper 3/4 surface are present.

| | | |
|---|---|---|
| Remarks | : | It is grows up to about 20 cm. TL and used as food by poor. |
| Importance | : | Fisheries of no interest. |
| Genus | : | Catla ( Cuvier and Vallenciennes ) 1844. |

Synopsis of the species of the genus Catla Cuvier and Vallenciennes dealt within this next.

## 9. *Catla catla* (Ham.) 1822

B. 3; D. 17-19 (3-4/14-16); P. 19-21; V. 9; A. 8 (3-5); C. 19; L.I. 40-43 L.tr. 7½/6½-9.

| | |
|---|---|
| Popular names | : Catla |
| Local names | : Tampra, Catla |
| Environment | : Benthopelagic, freshwater, brackish, depth range 5 m. |
| Climate | : Subtropical; 18-28°; 34 N - 195°. |
| Dangerous | : Harmless |

**Characters**

1. Length of head ranges from 4.1 to 4.7 and the height of body from 3 to 3.5 in the total length.

2. Body is deep and stout.

3. Mouth is wide and the lower jaw is prominent.

4. Barbels are absent.

5. Dorsal fin commences slightly in advance of the ventral fin.

6. Pectoral fin is located slightly behind the ventral fin.

7. Caudal fin is forked.

8. Scales are of the moderate size.

9. Lateral line is complete and commences from the upper margin of the gill cover.

10. Colour of the body is grayish above and silvery on the lateral and ventral sides fins are blackish in colour.

| | |
|---|---|
| Remark | : It is one of the most productive food fishes in India. It is one of the fastest growing fish in the country it grows up to 182 cm total length. |
| Importance | : Fisheries: highly commercial; aquaculture: commercial; game fish, yes. |
| Genus | : Cirrhinus (Oken) 1817. |

Synopsis of the species of the genus Cirrhinus Oken dealt within this text.

## 10. *Cirrhinus fulungee* (Sykes) 1841

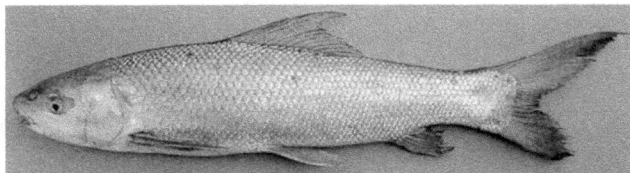

B. 3; D. 10 (2/8); P. 15; V. 9; A. 7 (2-5); C. 19; L.I. 48; L.tr. 8/9 Barbels 2.

Local names       : White Carp

Environment       : Benthopelagic, freshwater

Climate           : Tropical

Dangerous         : Harmless

**Characters**

1. Length of head is 5.7 and the height of body 5.2 in the total length.

2. Snout is slightly over hanging the mouth.

3. Mouth is broad and transverse.

4. Only one pair very short rostral barbels are present.

5. Dorsal fin commences midway between the end of the snout and the posterior end of the base of the anal fin.

6. Caudal fin is deeply forked.

7. Lateral line is complete.

8. Body is brownish along the back, divided by a light bluish band from the silvery abdomen. Dorsal and caudal fins are gray.

Remark            : It grows up to 30 cm in length.

Importance        : Fisheries; minor commercial.

## 11. *Cirrhinus mrigala* (Ham.) 1822

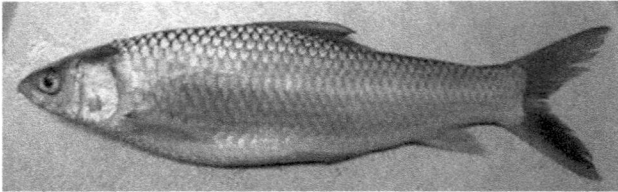

B. 3; D. 15-16 (3/12-13); P. 15-18; V. 9; A. 8(2/6); C. 15; L.I. 40- 45; L.tr 6½-7 / 6 ½- 8 ½; Barbels 2.

| | | |
|---|---|---|
| Popular names | : | Mrigal |
| Local names | : | Naim, mrigala |
| Environment | : | Benthopelagic, freshwater. |
| Climate | : | Tropical |
| Dangerous | : | Harmless |

**Characters**

1. Length of head ranges from 5 to 5.3 and the height of body from 4 to 5.5 in the total length.
2. Streamlined body, snout is blunt.
3. Mouth broad, upper lip complete, lower lip indistinct.
4. Single pair of barbels is present.
5. Dorsal fin as high as body or length of head.
6. Pectorals fin short, not reaching pelvic fin.
7. Caudal fin deeply forked.
8. Lateral line is complete.
9. Body is silvery, dark gray along the back, some times with coppery tinge.

| | | |
|---|---|---|
| Remarks | : | It attains the length of about 90 cm Weighting up to 8 kg in its naturals surroundings. It adults stage if feeds mainly on algae. Breeding takes place in flooded rivers during July-September. It is an excellent species for stocking the tanks. |
| Importance | : | Fisheries: commercial; aquaculture: Commercial game fish: yes. |

## 12. *Cirrhinus reba* (Ham.) 1822

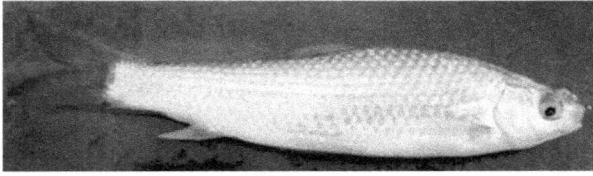

B. 3; D. 10-11 (2/3 | 8-9); P. 16; V. 9 A. 8 (2-3/5-6); C. 19; L.I. 35-38; L.tr. 7/5-9. Barbels 2.

| | | |
|---|---|---|
| Popular names | : | Mrigal, Mirrgah, Reba |
| Local names | : | Rewah, Raicheng |
| Environment | : | Benthopelagic, Freshwater |
| Climate | : | Tropical |
| Dangerous | : | Harmless |

**Characters**

1. Length of head ranges from 5.1 to 6.2 and the height of body from 4 to 5.2 in the total length.

2. Snout is slightly projecting.

3. Body is slightly more convex than the abdominal one.

4. Upper lip is fringed in young and adult forms.

5. Barbels are only two and rostral in position.

6. Dorsal fin originates anterior to the ventral fin.

7. Caudal fin is deeply forked.

8. Lateral line is complete.

9. Scales are hexagonal and there are five rows between the lateral line.

10. Colour of the body is silvery. Some times bluish longitudinal bands are found on body.

| | | |
|---|---|---|
| Remarks | : | It may grow up to a length of 30 cm. The male is smaller than the female. Breeding takes place in flooded marginal shallows during June-September. |
| Importance | : | Flesh is oily and tasteful and liked by the consumers. It is used as food in India. |

Genus               : Cyprinus

Synopsis of the species of the genus Cyprinus -------------dealt within this text.

### 13. *Cyprinus carpio carpio* (Linn.) 1758

B. 3; D. 3-4 /17-23; P. 19; V. 9; A. 2-3/ 5-6; C. 3/17-19 L.I. 35-37; Barbels 4.

Popular names      : Cyprinus, common carp, exotic carp.

Local names        : Cyprinus

Environment        : Benthopelagic, non migratory, freshwater, brackish, pH range 7- 7.5 ; dh range 10.0-15.0.

Climate            : Temperature ; 3 -32°C; 60°N-40°N.

Dangerous          : Potential pest

**Characters**

1. Body compressed and stout; head triangular.

2. Snout obtusely rounded, mouth oblique protrusible and small, lips fleshy.

3. Barbels two pairs rostral as long as maxillary.

4. Dorsal fin inserted at mid point of the length; dorsal spine stout serrated.

5. Anal fin trapezoidal in shape.

6. Pectoral fin large.

7. Caudal fin deeply emarginated.

8. Scales are large; lateral line scales 30 to 40.

9. There are three varieties:

    (a) *C. Carpio communis*: It is commonly called carp owing to the presence of regularly arranged rows of scales.

(b) *C. Carpio nudus*: It is commonly called leather carp owing to the general lack of scales.

(c) *C. Carpio specularis*: It is commonly called mirror carp it is characterized by the presence of few scales arranged in an irregular pattern or just scattered.

| | |
|---|---|
| Remarks | : It grows up to a length of 120 cm. SL. Maximum published weight 37.3 kg. And maximum reported age 47 years. |
| Importance | : Fishes; highly commercial; aquaculture; commercial; brackish; pH range 7.0-7.5; dH range 10.0, 15.0. |
| Genus | : Garra |

Synopsis of the species of the genus Garra------------dealt within this text.

## 14. *Garra lamta* (Ham.) 1822

B. 3; D. 11 (2-3/8-9); P. 15; V. 9; A. 7(2/5); C. 17; L.I. 32-36; L.tr. 4-4½ 5; Barbels 4.

| | |
|---|---|
| Popular names | : Stone sucker, Pathar cahta |
| Local names | : Pathar chata |
| Environment | : Benthopelagic, freshwater |
| Climate | : Tropical; 24-28°C |
| Dangerous | : Harmless |

## Characters

1. Length of head ranges from 5 to 5.5 and the height of body from 5 to 6 in the total length.

2. The eyes are directed slightly upwards and onwards and situated in the commencement of the last half of the head.

3. Body is elongated and subcylindrical.

4. Barbels four in number.

5. Dorsal fin arises midway between the end of the snout and the base of the caudal fin.

6. Caudal fin is slightly lobed.

7. Lateral line is complete.

8. Body is greenish brown or blue green above and yellowish below.

9. Generally a dark spot is present behind the gill opening.

Remark : This species is found in hill streams and is known to browse on algae covering rocking. It attains at least 14 cm in length.

Importance : Fisheries; commercial.

## 15. *Garra lissorhynchus* (McClelland, 1842)

B. 3; D. 10 (2/8); P. 15; V.9; A 6⅕; C. 19; L.I. 35; L.tr. 4½ I 3½; Barbels 4.

Popular names : Stone Sucker

Local names : Pathar chata

Environment : Benthopelagic, freshwater

Climate : Tropical; 24-28°C

Dangerous : Harmless

## Characters

1. Length of head ranges from 5.5 and the height of body from 5½ in the total length.

2. Head broad, depressed.

3. Body is elongated and subcylindrical.

4. Mouth, with an adhesive sucker, which is posterior to the lower jaw.

   5. Barbels four, one rostral and one maxillary pair.

   6. Dorsal fin arises in advance of the ventral.

   7. Caudal fin slightly forked.

   8. Colour of body is greenish brown, with no marks existing except of dark blotch under the dorsal fin.

Importance          : Fisheries of no importance.

Genus               : Labeo (Curvier) 1817.

   Synopsis of the species of the genus Labeo Cuvier dealt within this text.

## 16. *Labeo boga* (Ham.) 1822

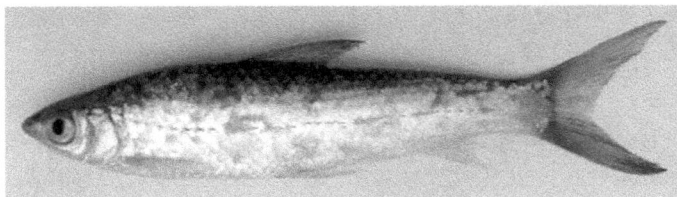

   B. 3; D. 11-13 (2-3/9-10); P. 16; V. 9; A. 7 (2/5); C. 19; L.I. 37-39; L.tr. 6½ - 7/7; Barbels 2.

Popular names       : Burmese fish, jumuna fish.

Local names         : Boga

Environment         : Benthopelagic, potamodromous, freshwater.

Climate             : Tropical

Dangerous           : Harmless

## Characters

   1. Length of head ranges from 5.2 to 5.5 and the height of body from 5.5 to 5.7 in the total length.

   2. Body is moderately elongated and the abdomen is rounded.

   3. Eyes are situated slightly before the middle of the length of the head.

   4. Barbels are two, minute maxillary in position.

   5. Dorsal fin commences in much advance of the ventral fins.

   6. Caudal fin is deeply forked with both of its equal lobes.

7. Lateral line is complete.

8. Body is orange with the fins of reddish tinge; sometimes a dark spot is present on the shoulder.

Remarks : It attains a maximum length of 30 cm; like other carp, it also spawn in flooded river.

Importance : It renowned as a tasty fish. This species can be culture in pond.

## 17. *Labeo boggut* (Skyes) 1841

B. 3; D. 11-12 (3/8-9); P. 17; V. 9; A. 7 (2/5); C. 19; L.I. 60-65; L.tr. 11-12 | 14. Barbels 2.

Popular names : Gubali, Boggut labeo.

Local names : Kolees

Environment : Benthopelagic, freshwater

Climate : Tropical

Dangerous : Harmless

## Characters

1. Length of head ranges from 5.5 to 6 and the height of body from 5.5 to 6.2 in the total length.

2. Body is moderately elongated and the abdomen is rounded.

3. Snout is thick slightly projecting beyond the jaws but without any lateral lobe.

4. Lower lip is fimbriated and internally lined with a horny covering.

5. Only one pair of short maxillary barbels is present.

6. Dorsal fin nearer to the snout than the root of caudal fin.

7. Pectoral is nearer about as long as the head.

8. The caudal fin is deeply forked.

9. Lateral line is complete and there are 8 to 9 rows of scales between it and base of ventral fin.

10. Body silvery, darkest dorsally but the fins are orange.

11. A dark spot present near the base of the caudal and sometimes a smaller one on the lateral line above the last third of the pectoral fin.

| | |
|---|---|
| Remarks | : It grows up to 20 cm in length. It is the most abundant freshwater fish in the inland waters of the Kathiwar and has good prospect for exploitation. It is also popular for stocking ponds. |
| Importance | : Fisheries: commercial; aquaculture: commercial. |

## 18. *Labeo Calbasu* (Ham.) 1822

B. 3; D. 16-18 (3/13-15); P. 19; V. 9; A. 7 (2/5); C. 19; L.I. 40-44; L.tr. 7½/8. Barbels 4.

| | |
|---|---|
| Popular names | : Kalbasu |
| Local names | : Kala-beinse; Kanoshi. |
| Environment | : Benthopelagic, Potamodromous, freshwater, brackish, depth range 10-11m. |
| Climate | : Tropical 25°N-16°N. |
| Dangerous | : Harmless |

## Characters

1. Length of head ranges from 5 to 6 and the height of body are 4 in the total length.

2. Body is moderately elongated and the abdomen is rounded.

3. Mouth is narrow; snout is obtuse and depressed. It is without lateral lobes but with pores.

4. One pair rostral and one pair of maxillary barbels are present, the rostral being slightly longer than the maxillary.

5. Dorsal fin commences midway between the snout and the base of the caudal fin.

6. Caudal fin is deeply forked.

7. Scales are large and 5½ to 6½ rows are present in between the lateral line and base of the ventral fin.

8. Lateral line is complete.

9. Body is blackish with scarlet in the center of the scales fins are black.

Remarks : It grows upto 91 cm in the length and is an excellent food fish. It is being stocked in the fish tanks at several places in India.

Importance : Fisheries: commercial, aquaculture: commercial.

## 19. *Labeo fimbratus* (Bloch) 1769

B. 3; D. 19-22 (3-4/15,18); P. 17; V. 9; A. 7(2/5); C. 19; L.I. 44-47; L.tr. 9-10/8 Barbels 4.

Popular names : Labeonain

Local names : Labeo

Environment : Benthopelagic, freshwater, Potamodromous

Climate : Tropical

Dangerous : Harmless

## Characters

1. Length of head ranges from 6.2 to 6.5 and the height of body from 4 to 4.5 in the total length.

2. Body is moderately elongated and the abdomen is rounded.

3. The snout is obtuse, rather swollen and beset with minute pores but it is divide of lateral lobes.

4. Two short rostral and two maxillary barbels are present.

5. The dorsal fin commences s little nearer the snout than to the base of the caudal fin.

6. Caudal fin deeply forked.

7. Lateral line is complete and there are 6 to 7 rows of scales between it and the base of the ventral fin.

8. Body is silvery along the back whitish below the fins are stained with black.

| | | |
|---|---|---|
| Remarks | : | This fish attains a maximum length of about 46 cm. It is good a good food fish. |
| Importance | : | Fisheries: commercial; aquaculture: commercial. |

### 20. *Labeo rohita* (Ham.) 1822

B. 3; D. 15-16 (3/12-13); P. 17; V. 9; A. 7 (2/5); C. 19; L.I. 40-42; L.tr. 6½- 7½/9 Barbels 2.

| | | |
|---|---|---|
| Populer names | : | Rohu |
| Local names | : | Rohu,Tambadamssa |
| Environment | : | Benthopelagic, freshwater, brackish,depth range 5 m. |
| Climate | : | Tropical 32°N-21°N |
| Dangerous | : | Harmless |

**Characters**

1. Length of head ranges from 4.5 to 5 and the height of body from 4 to 4.5 in the total length.

2. Body is moderately elongated and the abdomen is rounded.

3. The inter orbital space is flat.

4. Mouth small and inferior, lips thick and fringed, each lip with a distinct inner fold; snout depressed and projecting beyond mouth.

5. Only one pair of short and thin maxillary barbels is present.

6. Caudal fin is deeply forked.

7. The lateral line is complete.

8. The body is bluish black along the back; becoming reddish black along the sides and silvery beneath scales are with buff, orange or reddish center and dark margin. Fins are black.

Remarks : It is one of the most valuable food fishes of India and is very suitable for pond culture. It grows upto a length of 200 cm. It is estimated excellent as food and of great economic importance.

Importance : Fisheries: highly commercial; aquaculture: commercial game fish: yes.

Genus : Osteobrama (Heckel) 1842.

Synopsis of the species of the genus Osteobrama Heckel dealt within this text.

**21. *Osteobrama bakeri* (Day) 1873**

B. 3; D. 11 (3/8); P. 13; V. 10; A. 14 (3/11); L.I. 44; L.tr. 8/7. Barbels-4

Popular names : Malabar, osteobrama

Local names : Muchnee

Environment : Benthopelagic, freshwater

Climate : Tropical

Dangerous : Harmless

**Characters**

1. Length of head ranges from 5 to 6 and the height of body from 4 to 4½ in the total length.

2. Abdomen is rounded.

3. Mouth small, horseshoe shaped upper jaw the longer.

4. Barbels are four all very short.

5. Dorsal fin rather higher than body, and commencing, midway between the end of the snout and the base of the caudal fin.

6. Caudal deeply forked; lobes of equal length.

7. Scales are small; lateral line is complete and nearly straight.

8. Anal fin is long.

9. Colour of the body is silvery, caudal and dorsal edged with black.

Remarks            : It attain at least 11 cm in length.

Importance         : Fisheries no interest.

**22. *Osteobrama cotio cotio* (Ham.) 1822**

B. 3; D. 11-12 (3/4|8); P. 13-15; V. 10; A. 29-36 (2-3/27-33); C. 19; L.I. 55-70; L.tr. 9-15|14-21; Barbels-4

Local names        : Goordah, Chen-da-lah, Muchnee

Environment        : Benthopelagic, freshwater

Climate            : Tropical; 22-25°C

Dangerous          : Harmless

**Characters**

1. Length of head ranges from 5.5 to 6 and the height of body from 3 to 3.3 in the total length.

2. Body is very compressed.

3. Upper jaw is slightly longer than the lower jaw.

4. Barbels are absent.

5. Dorsal fin commences rather nearer the snout than the base of the caudal fin; its second osseous ray is weak and serrated.

6. Pectorals fin reaches over the origin of the ventral fin reaches the origin of the anal fin.

7. Caudal is deeply forked with the lower lobe being the longer.

8. Lateral line passes nearly to the center of the base of the caudal fin.

9. Scales are small.

10. Body is silvery; back is yellowish green with dull black dots.

Remarks : It grows upto a maximum length 15.0 cm. It is a useful larvicidal fish.

Importance : Fisheries of no interest.

Genus : Puntius (Hamilton) 1822

Synopsis of the species of the genus Puntius Hamilton dealt within this text.

## 23. *Puntius amphbius* (Valenciennes) 1842

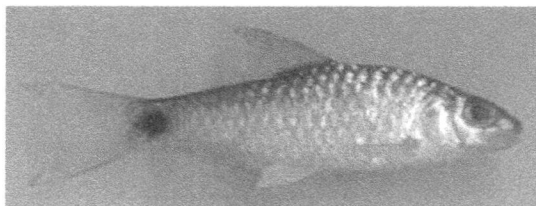

B. 3; D. 10-11; (2-3/8); P. 15; V. 9; A. 7 (2/5); C. 19; L.I. 23-24; L.tr. 5/4. Barbels 2.

Popular names : Scarlet - banded barb.

Local names : Khavali

Environment : Benthopelagic, freshwater, brackish

Climate : Tropical

Dangerous : Harmless

## Characters

1. Length of head ranges from 4.7 to 5 and the height of body from 4.2 to 5 in the total length.

2. Mouth is narrow with snout rounded and the upper jaw longer than the lower one.

3. Jaws are closely covered over by lips, rarely with leathery lobes.

4. Barbels are two maxillary thin and reaching to below the center of the eye.

5. The dorsal fin arises slightly in advance of the Ventrals and rather nearer.

6. Caudal fin is deeply forked.

7. Lateral line is complete and there are 2 rows of scales.

8. Body of the colour is steel blue, becoming white with golden tings at the side and below. Fins are yellowish, upper edge of dorsal tinged black.

Remarks             : It grows upto a maximum length 20 cm. The fish is locally consumed potentioalities as larvividal fish.

Importance          : Fisheries : subsistence fisheries.

**24. *Puntius chola* (Ham.) 1822**

B. 3; D. 11 (3/8); P. 14-15; V. 9; A. 7-8 (2-3/5); C. 19; L.I. 26 -28; L.tr. 5½ - 6½ I 5-5½. Barbels.

Popular names       : Green Barb

Local names         : Katcha, Karawa

Environment         : Benthopelagic,potamodromous, freshwater, brackish, pH- range 6-6.5 dH range 8-15.

Climate             : Tropical 20-25°C

Dangerous           : Harmless

**Characters**

1. Length of head ranges from 4.5 to 4.7 and the height of body from 3.2 to 3.7 in the total length.

2. Mouth is arched anterior or inferior.

3. Dorsal profile is more convex than that of the abdomen jaws are equal anteriorly.

4. Only one pair of maxillary barbels is present which are shorter than the orbit.

5. Dorsal fin originates opposite the ventral fin and is situated mid way between the ends of the snout.

6. Lateral line is complete.

7. Body is silvery operate shot with gold and purple.

8. Along the lateral line and located on 23rd and 24th scales is present a black blotch.

9. Another blotch is located on the base of the dorsal fin.

10. Occasionally there is a dark mark behind the gill opening.

Remarks : It is a small but useful species, which grows upto 13 cm in length. It is rather bitter in taste but eaten locally.

Importance : Fisheries: commercial. Aquaculture: commercial.

### 25. *Puntius Conchonius* (Ham.) 1822

B. 3; D. 11 (3/8); P. 11-15; V. 9; A. 7-8 (2-3/5); C. 19; L.I. 24-27; L.tr. 5½-6-6½.

Popular names : Khavli, Rosi barb.

Local names : Darahi

Environment : Benthopelagic, freshwater, pH range 6-8, dH range 5-19.

Climate : Subtropical 18-22°C; 40°N-8°N.

Dangerous : Harmless

## Characters

1. Length of head is 5 and the height of body ranges from 2.6 to 2.7 in the total length.

2. Body is evaluated and a slight concavity is there on the nape followed by a considerable rise to the base of the dorsal fin.

3. Barbels are absent.

4. The dorsal fin has its last undivided ray osseous, spinous and serrated.

5. The dorsal fin commences mid way between the anterior extremity or the orbit and the base of the caudal fin.

6. Caudal fin is deeply forked.

7. Lateral lines are complete, ceasing after 10 to 12 scales from its commencement.

8. Body is silvery dark along the back. All the scales are with black base.

9. A black blotch is situated on lateral line in area between 18th to 20th scale. Fins are transparent.

Remark                : It is attains up to 13 cm. In the length and is consumed locally by the poor classes of the people.

Importance            : Fisheries no interest, aquarium - commercial.

## 26. *Puntius filamentosus* (Valenciennes) 1844

B. 3; D. 11 (3/8); P. 15; V. 9; A. 7 (2/7); C. 19; L.I. 21; L.tr. 4-4½.

Popular names        : Filamented Barb

Local names          : Barb black spot

Environment          : Benthopelagic, freshwater, brackish, pH range 6-6.5, dH range 15

Climate              : Tropical 20-24°C

Dangerous            : Harmless

**Characters**

1. Length of head is 5 and the height of body ranges 3 to 3.5 in the total length.

2. Body is strongly compressed and elevated.

3. Snout is large pores are present.

4. Barbels are absent.

5. Dorsal fin commences rather nearer the snout than the base of the caudal fin.

6. Lateral line is complete and there are two rows of scales are situated before the dorsal fin.

7. Colour of the body is silvery.

8. A black mark is present near the posterior and of the lateral line before the base of the caudal fin.

9. All the fins are black.

Remark : It grows up to about 15 cm in length.

Importance : Fisheries: minor commercial aquarium: commercial bait: occasionally.

### 27. *Puntius jerdoni* (Day) 1870

B. 3; D. 12 (3/9); P. 15; V. 9; A. 8(3/5); C. 19; L.I. 27-28; L.tr. 6/4. Barbels 4

Popular names : Jerdon's carp

Local names : Potil

Environment : Benthopelagic, freshwater, potamodromous.

Climate : Tropical

Dangerous : Harmless

**Characters**

1. Length of head range from 5 to 5.2 and the height of body are 4 in the total length.

2. The head is nearly as high as it is long.

3. Body is laterally compressed.

4. Mouth is narrow and the upper Jaw is longer than the lower jaw.

5. Barbels are four, two are maxillary whish are as long as the orbit while the rostral ones are rather shorter.

6. The dorsal fin arises midway between the snout and the base of the caudal fin.

7. Pectoral fin is as long as the head.

8. Anal reaches rather beyond the root of the caudal fin.

9. Caudal fin is deeply forked.

10. Colour of the body is silvery fins are tinted orange and ipped with black.

Remarks               : This fish grows upto 46 cm in length. Suggesting economic possibilities.

Importance            : Fisheries: Commercial.

## 28. *Puntius sarana-sarana* **(Ham.) 1822**

B. 3; D. 11-12 (3/8-9); P. 15-16; V. 8-9; A. 8 (3/5); C. 19; L.I. 32-34 L.tr. 5½-6½|6; Barbels 4.

Popular name      : Olive Barb.

Local names       : Giddi-Kaoli, Durhie , Potah.

Environment       : Benthopelagic, freshwater, potamodromous, brackish

Climatc           : Tropical

Dangerous         : Harmless

**Characters**

1. Length of head ranges from 5 to 5.2 and the height of body from 3.5 to 3.7 in the total length.

2. Dorsal profile of the body is slightly elevated.

3. Interorbital space is convex.

4. Two pairs or barbels are present, rostral pair as long as the orbit maxillary pair is slightly longer.

5. Dorsal fin inserted nearer snout than caudal fin base.

6. Caudal fin is deeply forked.

7. Lateral line complete with 30 to 33 Scales.

8. Colour of the body is silvery, darkest dorsally, opercles golden fins dusky brown to orange, mostly some dark spots behind the opercle.

Remarks : It grows up to 31 cm in length and is much valued as food.

Importance : Fisheries: Minor commercial, game fish: yes, aquarium: commercial.

**29. *Puntius sophore* (Ham.) 1822**

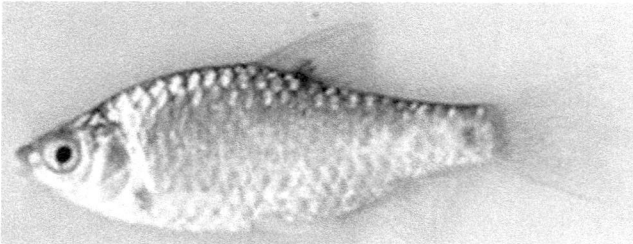

B. 3; D. 11-12 (3/8-9); P. 15-17; V. 9; A. 8 (3/5); C. 19 L.I. 23-26; L.tr. 4½-5½ I 5-5½.

Local names : Kateha -Karawa.

Environment : Benthopelagic, freshwater, brackish.

Climate : Tropical, 39°N-8°N.

Dangerous : Harmless

**Characters**

1. Length of head is 5 and the height of body ranges from 3.5 to .3.7 in the total length.

2. The dorsal profile is much more convex than abdomen and little concave over the occipit.

3. Upper jaw is slightly the longer.

4. Barbels are absent and the teeth are pharyngeal and crooked.

5. Dorsal fin inserted opposite the ventral fins.

6. The caudal fin is forked.

7. Lateral line is complete with 22 to 25 Scales.

8. Body is silvery with the dorsum darker and the opercles are shot with gold and red.

9. There are two black blotches, one at the base of the caudal fin along the lateral line over $22^{nd}$, $23^{rd}$ and $24^{th}$ Scales.

Remarks : It grows upto 18.0 cm in length. Poor people consume it in large quantities and it is larvicidal fish.

Importance : Aquarium : show aquarium.

### 30. *Puntius ticto* (Ham.) 1822

B. 3; D. 11 (3/8); P. 13-15; V. 9; A. 7-8 (2-3/5); C. 19; L.I. 23-26; L.tr. 5-6 | 6-6½

Popular names : Fire-fin Barb.

Local names : Kaoli

Environment : Benthopelagic, freshwater, brackish, pH range 6.5-7, dH range 10.0.

Climate : Tropical

Dangerous : Harmless

**Characters**

1. Length of head is 5 and the height of body ranges 3 to 3.2 in the total length.

2. Body is strongly compressed and elevated.

3. Upper jaw is slightly longer than the lower jaw.

4. Barbels are absent, teeth are pharyngeal, crooked and pointed.

5. Mouth is terminal and small.

6. Dorsal fin inserted slightly posterior to pelvic fin origin.

7. Caudal fin is forked.

8. Lateral line is complete with scales 23 to 25.

9. Colour in ornamental pattern showing iridescence and red edging of dorsal fin.

10. A black blotch is located on the side of the tail before the base of the caudal fin.

Remarks : It grows upto 10.0 cm in length. It is consumed as food by poors. It is also larvicidal fish.

Importance : Fisheries of no interest; aquarium: commercial.

Genus : Thynnichthyes (Bleeker) 1859.

Synopsis of the species of the genus Thynnichthyes Bleeker dealt within this text.

## 31. *Thynnichthyes sandkhol* (skyes) 1839

B. 3; D. 12 (3/9); P. 19; V. 9; A. 8 (3/5); C. 19; L.I. 120. L.tr. 25-30/25.

Popular names : Sandkhol Carp

Local names : Sandkhol

Environment : Benthopelagic, freshwater, potamodromous.

Climate              : Tropical

Dangerous            : Harmless

**Characters**

1. Length of head range from 4 to 4.5 and the height of body from 3.2 to 4 in the total length.

2. The eyes are situated in about the center of the depth of the head.

3. Abdomen is rounded; head is slightly compressed.

4. The lower jaw is little prominent upper lips absent.

5. Barbels are absent.

6. Dorsal fin aeries slightly in advance of the ventral and about midway between the snout and the base of the caudal fin.

7. Caudal fin is deeply forked; the lower lobe is longer.

8. Lateral line is complete there are 17 to 19 rows of scales.

9. Body is silvery and the head purplish in colour.

Remark          : It grows upto a maximum length of 61 cm. It is also very useful for stocking in temporary waters due to its fast growing quality.

Importance      : Fishenuies-commercias, aquaceutan ce commercial.

Genus           : Tor (Gray) 1833-34

Synopsis of the species of the genus Tor Gray dealt within this text.

**32. *Tor tor* (Ham.) 1822**

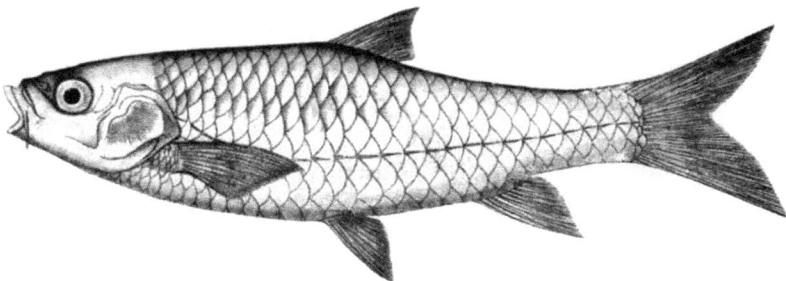

B. 3; D. 12 (3/9); P. 17-19; V. 9; A. 7-8(2-3/5); L.I. 25-27; L.tr. 4-4½ I 4-4½; Barbels 4.

| Local name | : Hindi, Mahasher, Naharm, Mar. Khadchi, Masta, Mahala. |
| Environment | : benthopelagic, freshwater, depth range 15cm. |
| Climate | : Tropical; 15-30°C; 29N-20N. |
| Dangerous | : Harmless |

**Characters**

1. Length of head ranges from 4 to 5.3 and the height of body from 4.3 to 5.5 in the total length.

2. Snout is rather pointed, jaws are of the equal length and the lips are thick.

3. Body is elongated and moderately compressed.

4. Mouth is inferior and strongly curved.

5. Barbels are two pair; maxillary pair is longer than rostral.

6. The dorsal fin arises opposite the ventral fin.

7. Caudal fin is deeply forked.

8. Lateral line is complete.

9. Scales are large.

10. Body is golden along the sides with dark gray colouration along the dorsal side.

11. Abdomen is silvery golden and the fins are reddish yellow.

| Remark | : It grows upto about 98 cm in length and ranks very high as food. Importance : Used as tasty food fish. This species can be utilized in biological control of water weeds (Tilak and Sharma, 1982). It is a sporting fish and also very profitable for angler (Talwar and Jhingran, 1991). This species command a good market price and consumer demand. |
| Subfamily | : Hypothalamichthyine. |
| Genus | : Hypothalamichthys Bleeker (1859). |

Synopsis of the species of the genus Hypothalamichthys Bleeker dealt within this text.

### 33. *Hypothalamichthys molitrix* (Valenciennes) 1844

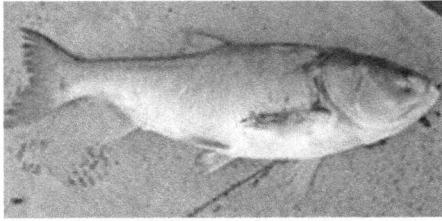

D. 6-7 (3/3-4); P. 10; A. 10-14 (3/7-11); V. 9; C. 18; L.I. 110-115.

| | |
|---|---|
| Local names | : Silver carp |
| Environment | : Benthopelagic, Potamodromous, freshwater, depth range 0-20 m. |
| Climate | : Temperate - 6-28°C, 64°N-43°S. |
| Dangerous | : Potential pest |

**Characters**

1. Body stout, compressed, Abdomen strongly compressed with a sharp keel from breast to vent.

2. Head moderate, snout bluntly round.

3. Mouth anterior, large, wide, cleft not extending to anterior margin of eye.

4. Eyes rather small, anterior, sub inferior, visible from below ventral surface.

5. Upper jaw a little protruded upward a little longer than the lower.

6. Barbels are absent.

7. Keels extend from isthmus to anus.

8. Dorsal fin inserted behind pelvic fins, or above lip of pectorals fins with 10 rays.

9. Anal fin with 14-17 rays (12-14 branched).

10. Caudal fin is forked.

11. Scales minute, lateral line decurved, continuous with 110-115 scales.

| | |
|---|---|
| Remarks | : It grows up to 105 cm in length. |
| Importance | : Fisheries: commercial; aquaculture: commercial. |

Genus                    :  Ctenopharyngodon (Valenciennes, 1844).

Synopsis of the species of the genus Ctenopharyngodon------------ dealt within this text.

## 34. *Ctenopharyngodon idella* (Valenciennes) 1844

Popular names       :  Grass Carp

Local names          :  Gavtaya, grass carp

Environment         :  Demersal, potamodromous, freshwater, depth range 0-30 m.

Climate               :  Temperate, 0-35°C, 65°N - 25°N.

## Characters

1. Body moderately elongate, subcylindrical anteriorly and compressed posteriorly.

2. Abdomen rounded; head depressed and flattened; Snout obtusely rounded.

3. Mouth terminal, cleft not extending to anterior margin of eyes.

4. Eyes large, lateral, in anterior part of head, may or may not be just visible from below ventral surface.

5. Upper jaw slightly longer than lower jaw and protractile.

6. Barbels are absent.

7. Dorsal fin inserted slightly ahead of pelvics nearer tip of snout than caudal base, with 10 rays 7 branched and without a spine.

8. Caudal fin well forked.

9. Anal fin short, with 10 rays, eight (8) branched.

10. Scales large, cycloid; lateral line continuous, slightly decurved, with 40-42 scales.

| Remarks | : It grows upto 150 cm in the length. |
|---|---|
| Importance | : Fisheries: minor commercial; aquaculture: commercial, game fish: yes. |
| Subfamily | : RASBORINAE |
| Genus | : Barilius Hamilton 1822 |

Synopsis of the species of the genus Barilius Hamilton dealt within this text.

### 35. *Barilius barila* (Ham.) 1822

B. 3; D. 9 (2/7); P. 13; V. 9; A. 13-14 (3/10-11); C. 19; L.I. 43-46; L.tr. 7/5 Barbels 4.

| Local names | : Hindi-presee |
|---|---|
| Environment | : benthopelagic, freshwater |
| Climate | : Tropical |
| Dangerous | : Harmless |

### Characters

1. Length of head ranges from 5 to 5.5 and the height of body from 5.2 to 5.5 in the total length.

2. Abdomen is rounded; Jaws are compressed.

3. A small rostral pairs of barbels are present.

4. Dorsal fin arises midway between the posterior margin of the orbit and the base of the caudal fin.

5. Pectoral fin is nearly as long as head.

6. Caudal is forked with the lower lobe slightly longer.

7. Scales are of moderate or small size.

8. Lateral line is complete, incomplete or even absent.

9. Body is silvery in colouration, with 14 or 15 vertical blue bands on each side.

| Remarks | : This fish grows up to 10 cm in length and relished as food by the poor class. |
|---|---|
| Importance | : Fisheries: of no interest; bait: usually. |

## 36. *Barilius barna* (Ham.) 1822

B. 3; D. 9 (2/7); P. 15; V. 9; A. 13-14 (3/10-11); C. 19; L.I. 39-42, L.tr. 8-916.

| | |
|---|---|
| Local names | : Dudhnia, Boroni |
| Environment | : Benthopelagic, freshwater |
| Climate | : Tropical |
| Dangerous | : Harmless |

**Characters**

1. Length of head ranges from 4.7 to 5.2 and the height of body from 3.5 to 4 in the total length.

2. Abdomen is rounded.

3. Open pores are present on both the Jaws and snout in adults.

4. Barbels are absent.

5. The pectoral fins reach above the ventral.

6. Anal commences under the middle or at the end of the dorsal fin.

7. Scales are of moderate or small size.

8. Lateral line is complete.

9. Colour of the body is dull green with 9 to 11 vertical dark bands on the sides.

10. The edges of the dorsal and caudal fins are black.

| | |
|---|---|
| Remarks | : It attains 12 cm or more in length and constitutes of the poor men's dish. |
| Importance | : Fisheries of no interest. |

## 37. *Barilius bendelisis* (Ham.) 1807

B. 3; D. 9 (2/7); P. 15; V.9; A. 9-10 (2-3 /7-8); C. 18; L.I. 40-43; L.tr. 7-8/5; Barbels 4.

| | |
|---|---|
| Local names | : Johra |
| Environment | : Benthopelagic, freshwater |
| Climate | : Tropical |
| Dangerous | : Harmless |

**Characters**

1. Length of head ranges from 4.7 to 5.2 and the height of body from 4.2 to 5.2 in the total length.

2. The maxilla reaches to bellow the first third or the orbit.

3. Four Barbels are present, rostral pair occasionally absent.

4. Dorsal fin commences nearer the base of the caudal fin than the snout.

5. Ventral may or may not reach the vent.

6. Caudal fin is forked; its lower lobe is usually longer.

7. Scales are small size.

8. Lateral line present.

9. Colour of the body is slivery, shot with purple and with short vertical bars on the dorso-lateral sides of the body. Fins are whitish, tinged with orange.

| | |
|---|---|
| Remarks | : It attains at least 15 cm in length. |
| Importance | : Fisheries of no interest bait, Usually. |
| Genus | : Rasbora Bleeker 1860. |

Synopsis of the species of the genus Rasbora Bleeker dealt within this text.

## 38. *Rasbora daniconius* (Ham.) 1822

B. 3; D. 9 (2/7); P. 15; V. 9; A. 7 (2/5); C. 19; L.I. 31-34; L.tr. 4½-5½ 4½-5.

| | |
|---|---|
| Popular name | : Common Rasbora |
| Local names | : Dandai, Dondai gana |
| Environment | : Benthopelagic, freshwater, Potamodromous, brackish pH range 7, dH range 12. |
| Climate | : Tropical 24-28°C |
| Dangerous | : Harmless |

**Characters**

1. Length of head ranges from 4.5 to 5 and the height of body from 4.5 to 5.2 in the total length.

2. Body is cylindrical and slightly compressed with rounded abdomen.

3. Lower jaw is with a central and two lateral prominences.

4. Barbels are absent.

5. Dorsal fin arises nearer to the origin of the ventral than that of the anal.

6. Caudal fin is forked.

7. Lateral line is concave.

8. Body is greenish yellow in the upper half and is shot with black dots. Lower half of the body is silvery white. The tips of the fins are brownish black or deep yellow.

| | |
|---|---|
| Remarks | : It attains a maximum length of 14 cm. It is an efficient larvicidal fish and consumed as food by the poor classes of people. |
| Importance | : Fisheries : minor commercial; aquarium : commercial. |

Family                 : Cobitidae

Genus                  : Lepidocephalichthys Bleeker 1863.

Synopsis of the species of the genus Lepidocephalichthys Bleeker dealt within this text.

### 39. *Lepidocephalichthys guntea* (Ham.) 1822

B. 3; D. 8-9 (2/6-7); P. 8; V. 7-8; A. 7 (2/5); C. 16; L.tr. 115; Barbels 8.

Local names            : Gunguch

Environment            : Demersal, Freshwater

Climate                : Tropical

Dangerous              : Harmless

## Characters

1. Length of head ranges from 6.5 to 6.7 and height of body from 5.7 to 6.5 in the total length.

2. Body is elongated, moderately compressed and the back is not elevated.

3. There are two pairs of rostral and one pair of maxillary barbels, all are longer than the orbit.

4. Dorsal fin is situated opposite the ventral and the caudal fin is entire.

5. Scales are minute; lateral line is absent.

6. Colour of the body is dirty yellowish brown.

7. A black band runs from the snout to the tail region.

8. A black ocellus is present above the middle of the base or the caudal fin, placed just above the lateral band.

9. The back is dotted black and the caudal and the dorsal fins are with numerous rows of darks spots.

Remark                 : It reaches a maximum length of 24 cm.

Importance             : Fisheries of no interest.

Genus                  : Nemacheilus Van Hasselt, 1823.

Synopsis of the species of the genus Nemacheilus Van Hasselt, Dealt within this text.

**40.** *Nemacheilus aureus* **(Ham.) 1822**

B. 3; D. 11-12 (2/9-10); P. 11; V. 8; A. 7 (2/5); C. 17; Barbels 6.

Local names        : Nemi.

Environment        : demersal, freshwater.

Climate            : Tropical

Dangerous          : Harmless

**Characters**

1. Body is elongated and the dorsal profile is nearly horizontal.

2. Barbels are six, long and the maxillary pair reaches to below the orbit.

3. Dorsal fin arises slightly nearer to the snout.

4. Pectoral is as long as the head.

5. Ventral is inserted under the middle of the dorsal fin.

6. Lateral line ceases opposite the posterior end of the dorsal fin.

7. Scales are indistinct.

8. Body is grayish, with from 10 to 14 short bars on the lateral line

9. Dorsal fin orange and with rows of block spots.

Remarks            : It attains at least 11.0 cm in length.

Importance         : Fisheries of no interest.

**41.** *Nemacheilus beavani* **(Ham.) 1822**

B. 3; D. 10 (2/8); P. 11; V. 7; A. 7 (2/5); C. 19. Barbels 6.

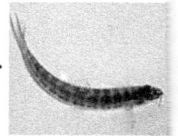

Local names        : Creek loach

Environment        : Demersal, freshwater

Climate            : Tropical

Dangerous          : Harmless

**Characters**

1. Length of head ranges from 5 and height of body from 5½ to 5¾ in the total length.

2. Eyes small, just before the middle of the length of the head.

3. Barbels six, four rostral and two maxillary.

4. Dorsal fin with an oblique upper edge, it arises slightly nearer the end of the snout then the base of the caudal.

5. Pectoral extends ⅔ of the distance to the root of the ventral.

6. Caudal fin is lobed.

7. Scales are minute.

8. Colour of the body is black streak, body with nine dark cross bands. Dorsal and caudal rays with blackish dots.

Remarks : It attains upto 10.2 cm in length.

Importance : Fisheries of no interest.

### 42. *Nemacheilus botia* (Ham.) 1822

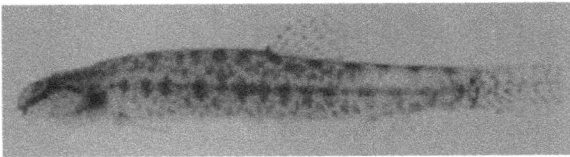

B. 3; D. 12-14 (2/10-12); P. 11; V. 8; A. 7 (2/5); C. 17. Barbels 6.

Popular names : Striped Loach

Local names : Botia

Environment : Demersal, freshwater

Climate : Tropical

Dangerous : Harmless

**Characters**

1. Length of head ranges 4.5 to 5.5 and the height of body are 4.7 in the total length.

2. Body is elongated and dorsal profile in early horizontal.

3. Preorbital has moveable projection below the orbit.

4. Barbels are six, long and maxillary pair reaches to below the hind border of the eyes.

5. Dorsal fin aeries slightly nearer to the snout than to the base of caudal fin.

6. Pectoral is as long as the head.

7. The caudal is slightly notched.

8. The scales are indistinct.

9. Lateral line is complete and there are 12 rows of scales.

10. Body is colour is grayish with 10-14 short vertical bars on the lateral line and a number of blotches above.

11. Dorsal fin is orange with rows of black spot.

Remarks          : It attains at least 7.6 cm in length.

Importance       : Fisheries of no interest.

## DIVISION - SILURI

Suborder         - Siluroidei

Family           - Bagridage

Genus            - Mystus Gronow

Synopsis of the species of the genus Mystus Gronow dealt within this text.

### 43. *Mystus aor* (Ham.) 1822

B. 12; D. 8 (1/7) 10; P. 10-11 (1/9-10); V. 6; A. 12-13 (3-4/9) C. 17; Barbels 8.

Popular names    : Mystus

Local names      : Katarna

Environment      : Dermisal, freshwater

Climate          : Tropical

Dangerous        : Harmless

## Characters

1. Length of head ranges from 4.5 to 5 and the height of body from 6 to 8.8 in the total length.

2. Body is elongated and naked mouth is terminal and transverse.

3. Snout is broad and spatulate.

4. Mouth is sub terminal.

5. The upper jaw is longer.

6. Barbels are eight, maxillary barbells reaches the base of the caudal fin or even beyond it.

7. Dorsal spine has its posterior edge finely serrated.

8. Adipose fin long, having its base twice as that of the rayed dorsal fin.

9. A dark spot on the tip of the adipose fin.

10. Body is bluish leaden above becoming white on abdomen fins are yellowish, tinted gray.

Remarks　　　　　: It is said to attain upto 180 cm in length this fish is eaten locally.

Importance　　　　: Fisheries-commercial, Game fish: yes.

## 44. *Mystus bleekeri* (Day) 1878

B. 10; D. 8 (1/7) 10; P. 10-11 (1/9-10); V. 6; A. 9-10 (3/6-7); C. 17; Barbels 8.

Popular names　　: Mystus

Local names　　　: Tengara

Environment　　　: Demersal, freshwater, pH range 5.8-8.0 dH range 30.0.

Climate　　　　　: Tropical

Dangerous　　　　: Harmless

## Characters

1. Length of head ranges from 5 to 5.5 and the height of body from 5 to 5.2 in the total length.

2. Body is elongated and naked mouth is terminal.

3. Occipital process twice as long as broad at its base, and extends the basal bone of dorsal fin.

4. Median longitudinal grooves shallow, reaching base of occipital process.

5. Barbels eight, maxillary extends to anal fin.

6. Dorsal spine is smooth and equal to half the length of the head.

7. Pectoral spine is serrated and stronger than the dorsal.

8. Adipose dorsal originates just behind the rayed dorsal with its base twice the head length.

9. Caudal fin is pointed lobes is forked, the upper lobe being longer than the lower.

10. Body is brownish gray with two light longitudinal bonds above and below the lateral line.

11. A dark shoulder spot is found on either side below the lateral line.

Remarks         : It attains at least 15.5 cm in length.

Importance      : Fisheries: minor commercial; aquarium: commercial.

## 45. *Mystus cavasius* (Ham.) 1822

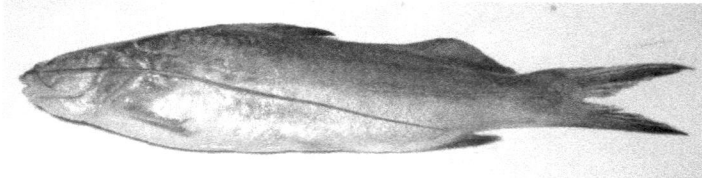

B. 6; D. 8 (1/7) I 0; P. 9-10 (1/8 -9); V. 6; A. 11-13 (3-4 I 7-9); C. 16; Barbels 8.

Local names       : Tengara

Environment      : Demersal, freshwater, brackish

Climate            : Tropical, 5°N-38°N

Dangerous        : Venomous

## Characters

1. Length of head ranges from 5.5 to 6.2 and the height of body from 5.2 to 6.5 in the total length.

2. Body is elongated and naked mouth is terminal and transverse.

3. Snout is somewhat obtuse, upper jaw is the longer.

4. Cleft of mouth extends to below the orbit.

5. Head conical, median longitudinal groove extending to base of occipital process.

6. Barbels are eight, the maxillary barbels extends beyond the base of the caudal fin.

7. Dorsal spine weak, feebly serrated.

8. Adipose fin large, inserted close behind the dorsal fin.

9. Pelvic fin originated just behind the base of the dorsal fin.

10. Caudal fin is forked, the upper lobe being the longer and more pointed.

11. Body is leaden above and yellowish on the abdomen and checks.

12. There is a black spot covering the basal bone of the dorsal fin, pectoral, pelvic and anal fins are adult white.

Remarks               : It attains about 40 cm in length. It is predatory in habit and attacks small carps, other small teleosts and prawns.

Importance            : Fisheries commercial

## 46. *Mystus seenghala* (Sykes) 1839

B. 12; D. 8 (1/7)|0; P. 10 (1/9); V. 6; A. 11-12 (3/8-9); C. 19-21; Barbels 8.

Local names           : Ari, Pogal, Singala, Singata.

Environment           : Demersal, freshwater, brackish.

Climate               : Tropical, 39°N- 8°N

Dangerous             : Harmless

## Characters

1. Length of head ranges from 4.1 to 4.5 and the height of body from 7.5 to 8.0 in the total length.

2. Clef mouth is shallow.

3. Body is elongated and necked, mouth is sub terminal and transverse.

4. Snout board and spatulate.

5. Upper jaw is longer than the lower.

6. Barbels are 8, maxillary pair reaching to middle or just beyond rayed dorsal but never reaches the caudal fin.

7. Dorsal spine weakly serrated on posterior edge.

8. Adipose fin base short, as long as ranged dorsal fin base.

9. Pectoral spine is stronger than the dorsal and serrated.

10. Caudal fin is deeply forked and its upper lobe longer.

11. Body is brownish gray superiorly and silvery on sides and abdomen.

12. A black spot is found on the hind end of the base of the adipose fin.

Reamrks : It attains at least 98 cm in length it is predatory in habbit and attcks small carps other small teteosts and prawns.

Importance : Fisheries, commercial, game fish yes.

## 47. *Mystus tengara* (Ham.) 1822

B. 10; D. 8 (1/7)I0; P. 9 (1/8); V. 6; A. 11-13 (2-3/9-10); C. 9; Barbels 8.

Local names : Tengara.

Environment : Demersal, freshwater.

Climate : Tropical

Dangerous : Harmless

**Characters**

1. Length of head ranges from 4.7 and the height of body from 4.5 in the total length.

2. Body is elongated and naked, mouth is terminal.

3. Upper jaw slightly longer.

4. Barbels are eight the maxillary barbels extend beyond the base of the ventral fin.

5. Dorsal fin as long as the head excluding the snout, slightly serrated.

6. Pectoral spine nearly as long as the stronger than that of the dorsal fin.

7. Caudal fin is forked, upper lobe the longer.

8. Colour of the body is brilliant yellow with a black shoulder spot and about five black longitudinal lines.

| | | |
|---|---|---|
| Remarks | : | It grows upto 18cm in total length. |
| Importance | : | Used as food fish in Bangladesh. It is very good to taste and useful in calcium deficiency (Bhuiyan, 1964). |

### 48. *Mystus Vittatus* (Bloch) 1794

B. 10; D. 8 (1/17) 10; P. 9-10 (1/8-9); V. 6; A. 9-12 (2-3/8-9); C. 17; Barbels 8.

| | | |
|---|---|---|
| Local names | : | Kala tengra |
| Environment | : | Dermisal, freshwater, brackish, pH range 6.0 -7.5, dH range 4.0-25.0. |
| Climate | : | Tropical, 22-28°C; 38°N-0°N |
| Dangerous | : | Harmless |

**Characters**

1. Length of head ranges from 4.5 to 5 and the height of body from 4.5 to 5.5 in the total length.

2. Body is elongated and mouth is terminal.

3. Median longitudinal groove on the head short, not extending to base occipital process.

4. Occipital process 3 times as long as broad at its base, reaching basal bone of dorsal fin.

5. Barbels are 8 the maxillary beyond pelvic fins, upto the anal fin end.

6. Dorsal spine is finely serrated.

7. Pectoral spine is also serrated and as long as head excluding snout.

8. Adipose fin small, inserted much behind rayed dorsal fin in advanced of anal fin.

9. Body is golden with dark bluish shoulder spot.

10. A broad black longitudinal band extends along and on either side of the lateral line. Another band is found along the dorsal side of the body.

| | | |
|---|---|---|
| Remarks | : | It attains about 20 cm in length. This fish is esteemed as food for its pleasant Smokey flavour. |
| Importance | : | Fisheries of no interest, aquarium commercial. |
| Genus | : | Rita (Bleeker) 1853. |

Synopsis of the species of the genus Rita Bleeker dealt within this next.

## 49. *Rita rita* (Ham.) 1822

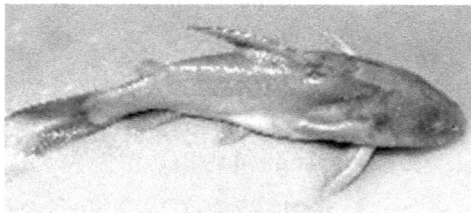

B. 8; D. 7; (1/6) I 0; P. 11 (1/10); V. 8; A. 13-14(4-5/9); C. 19; Barbels 6.

| | | |
|---|---|---|
| Local names | : | Rita |
| Environment | : | Demersal, freshwater, brackish, pH range 6.5-8.0, dH range 30-0. |
| Climate | : | Tropical, 18-26°C. |
| Dangerous | : | Harmless |

**Characters**

1. Length of head ranges from 4 to 4.3 and the height of body from 5.5 to 6 in the total length.

2. Mouth is transverse and upper jaw is the longer.

3. Head depressed occipital process subcutaneous extends to predorsal plate.

4. Barbels are six, and nasal very short, maxillary extends to operculum.

5. Dorsal spine is very strong, serrated.

6. Pectoral spine is shorter than the dorsal and is denticulated on both edges.

7. Caudal fin is forked.

8. Body is greenish gray above becoming lighter below.

Remark : It grows upto on 150 cm in the total length.

It retains its vitality for a long time after it is taken out of water and may be solid alive in local markets.

Importance : Fisheries commercial

Family : CLARIDAE (Air breathing catfishes)

Genus : Clarias Gronovius (emend. scopoli) 1763

Synopsis of the species of the genus Clarias Gronovius dealt within this text.

**50. *Clarias batrachus* (Linnaeus) 1758**

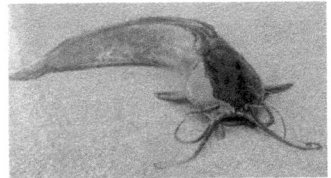

B. 9; D. 62-76; P. 9-12; (1/8-11); V. 6; A. 48-58; C. 15-17; Barbels. 6

Local names : Mangur, Magur, Mangri.

Environment : Dermisal, potomodromous, freshwater, brackish, depth range 1 m.

Climate : Tropical, 18-26°C

Dangerous : Potential

**Characters**

1. Length of head is 5.6 and the height of is from 6.5 to 7.5 in the total length.

2. Head is depressed and the body is elongated.

3. Gape of mouth is moderate, anterior and transverse.

4. Occipital process angular and narrow.

5. Barbels are eight; maxillary extends beyond base of pectoral fin.

6. Dorsal fin is very long; it originates a little behind the occipital process and ends a bit anterior to the base of the caudal fin.

7. Caudal fin is free.

8. Body is brownish black.

9. All the fin or covered with thick skin.

Remarks : It attains a maximum length of 47.0 cm. Due to the presence of accessory respiratory organs, it is useful for culture in swamps. It is supposed to be very nourishing fish.

Importance : Fisheries: commercial; aquaculture : commercial aquarium: commercial.

### 51. *Clarias gariepinus* (Burchell) 1822

B. 9; D. 61-80; P. 1/7│0; V. 6; A. 45-65; C. 17; Barbels 8.

Local names : Mangur

Environment : Benthopelagic, potomodromous, freshwater, pH range 6.5-8,dH range 5-28, depth range 4-8 m.

Climate : Subtropical, 8-35°C, 52°N-28°N

Dangerous : Potential pest

**Characters**

1. Body is elongated and the head is depressed.

2. Mouth is moderate, anterior and transverse.

3. Barbels are eight in number; maxillary reach the end of the pectoral fin, the mandibular ones are shorter.

4. Pectoral extends nearly to below the origin of the dorsal fin, its spine moderately strong.

5. Pectoral fin spine serrated only on its outer side.

6. Caudal fin is free.

7. Body colour generally dark, grayish black above creamy white below, a fairly distinct black longitudinal band on each side of the ventral side of the head.

| | |
|---|---|
| Remarks | : Occurs mainly in quite waters, lakes and pools but may also occur in fast flowing rivers. It grows upto 170 cm in length. |
| Importance | : Fisheries :minor commercial; aquaculture: commercial; game fish : yes. |
| Family | : Heteropnestidae |
| Genus | : Heteropnestes |

Synopsis of the species of the genus Heteropnestes Muller dealt within this text.

## 52. *Heteropnestes fossils* (Bloch) 1785

B. 7; D. 6-7; P. 8 (1|7); V. 6; A. 60-79; C. 19; Barbels 8.

| | |
|---|---|
| Local names | : Singi, Bitchuka, machi. |
| Environment | : Demersal, freshwater, brackish. |
| Climate | : Tropical |
| Dangerous | : Harmless |

**Characters**

1. Length of head ranges from 5.5 to 7 and the height of body from 5 to 8 in the total length.

2. Body is subcylindrical upto pelvic, compressed behind.

3. Head depressed occipital process not extending to base of dorsal fin.

4. Mouth small terminal.

5. Barbels are eight, maxillaries reaching middle of the pectoral or even to the pelvic fin.

6. Dorsal fin short spineless.

7. Ventral fin originates below the dorsal fin origin.

8. Pectoral spine is serrated.

9. Anal fin is long based, with 60-79 fins rays, not united with caudal.

10. The body is dark leaden brown; the young are reddish.

| | | |
|---|---|---|
| Remark | : | It attain 30 cm in the length or more. The pectoral spine of this fish is capable of inflicting wound which is very painful, and hence people are afraid of catching it without breaking the pectoral spine It is considered to be very nourishing and tasty fish. |
| Importance | : | Fisheries: highly commercial; aquaculture: commercial; aquarium: commercial. |
| Family | : | SCHILBEIDAE |
| Genus | : | Eutropiichthys Bleeker 1822 |

Synopsis of the species of the genus Eutropiichthys Bleeker dealt within this text.

**53.** *Eutropiichthys vacha* **(Ham.) 1822**

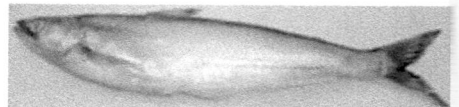

| | | |
|---|---|---|
| Local names | : | Bachwa, sugwaba. |
| Environment | : | Pelagic, potomodromous, freshwater, brackish. |
| Climate | : | Tropical |
| Dangerous | : | Harmless |

**Characters**

1. Length of head ranges from 5.5 to 5.7 and the height of body from 5 to 5.5 in the total length.

2. Body is elongated and compressed, head is covered with soft skin.

3. Eyes are with broad adipose lids.

4. Cleft of mouth is oblique reaching beyond the midorbit.

5. The snout is compressed and pointed; upper jaw is slightly longer.

6. Barbels are 8, maxillary reaching nearly the base of the pectoral fin.

7. The dorsal spine is thin serrated and as long as the head without snout.

8. Pectoral spine is serrated and slightly longer than the dorsal spine.

9. Anal fin is long with 44 to 51 rays.

10. Caudal fin is deeply forked.

11. Body is slivery grayish along the back. Edges of the pectoral and caudal fins are black.

Remark          : It attain a length of about 30 cm and good for eating.

Importance      : Fisheries: commercial; game fish : yes.

Genus           : Pangasius Valenciennes 1840.

Synopsis of the species of the genus pangasius Vatenciennes within this text.

## 54. *Pangasius pangasius* (Ham.) 1822

B. 9-10; D. 8 (1|7)|0; P. 13 (1/12); V. 6; A. 31-34 (4-5/27-29); C. 19; Barbels 9.

| | | |
|---|---|---|
| Local names | : | Parisasi, Pangsa |
| Environment | : | Benthopelagic, potomodromous, freshwater, brackish, pH range 6-7.5; dH range 25; depth range 50 m. |
| Climate | : | Tropical, 23-28°C; 35°N-8°N. |
| Dangerous | : | Harmless |

**Characters**

1. Length of head ranges from 5.5 to 6 and the height of body from 4 to 5 in the total length.

2. Body is elongated and compressed, head is slightly granulated above.

3. Eyes are situated in the anterior half of the head.

4. Cleft of mouth reaches opposite the center of front edge of the eye.

5. Barbels are four, maxillary ones extending to the base of the pectoral fin.

6. Dorsal spine is serrated, moderately strong.

7. Pectoral spine is serrated, strong as long as the dorsal spine.

8. Caudal is forked, lobes not sharply pointed.

9. Body is silvery, darkest superiorly; shot with purple on sides, cheeks and under the surface of the head is golden.

| | | |
|---|---|---|
| Remarks | : | It attains 122 cm in length. It also feeds on young carp fishes and molluscs. |
| Importance | : | Fisheries: commercial; aquaculture: Commercial Game fish: yes. |
| Genus | : | Pseudeutropius Sykes, 1839. |

Synopsis of the species of the genus Pseudeutropius Bleeker dealt within this text.

### 55. *Pseudeutropius taakree* (Day) 1839

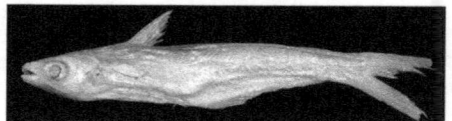

B. 6; D. 13-14 (1/6-7)|0; P. 1/10-11; V. 6; A. 43-52(3-4|40-50), C. 17. Barbels 8.

| | | |
|---|---|---|
| Local names | : | Moonia, Munvi |
| Environment | : | Freshwater; demersal |
| Climate | : | Tropical |
| Dangerous | : | Harmless |

## Characters

1. Length of head ranges from 6 to 6.5 and the height of body from 5 to 6 in the total length.
2. Body is elongated and compressed, head covered with soft skin.
3. The Cleft of mouth extended to opposite the middle of the front edge of the eye.
4. Barbels are eight.
5. Dorsal spine smooth anteriorly, serrated posteriorly as long as the head.
6. Pectoral spine stronger, serrated, internally and almost as long as the head.
7. Body of the colour is silvery, with a gloss of green along the black, Caudal stained with gray at its edges.

| | | |
|---|---|---|
| Remarks | : | It attains 19cm in length and good eating. |
| Importance | : | Fisheries: Minor Commercial. |
| Family | : | SISORIDAE |
| Genus | : | Bagarius (Bleeker) 1853 |

Synopsis of the species of the genus Bagarius Bleeker dealt within this text.

## 56. *Bagarius bagarius* (Ham.) 1822

B. 12; D. 7 (1/6)|0; P. 13 (1/12); V. 6; A. 13-15 (3/10-12); C. 17; Barbels 8.

Local names     : Khrit

Environment     : Benthopelagic, potomodromous, freshwater, brackish.

Climate     : Tropical, 18-25°C.

Dangerous     : Harmless

**Characters**

1. Length of head 4.5 to 4.6 and the height of body range from 8 in the total length including the caudal filament.

2. Body is 19 cm, depth 5.6 to 7.2 times in standard length.

3. Head depressed, its upper surface osseous.

4. Mouth is inferior and concentric.

5. Barbels are eight one nasal, one maxillary, and two mandibular pairs.

6. Dorsal inserted nearer adipose fin than to snout tip.

7. Pectoral serrated stronger as long as the dorsal fin.

8. Caudal fin is forked its upper lobe prolonged.

9. Adipose fin small.

10. Colour of the body is gray or yellowish, with large, irregular brown or black markings and cross bands.

Remarks     : It attains in the length.

Importance     : Fisheries: minor commercial; game fish: yes.

Family     : SILURIDAE

Genus     : Ompak (Lacepede) 1803

Synopsis of the species of the genus Ompak Lacepede dealt within this text.

**57. *Ompak bimaculatus* (Bloch) 1794**

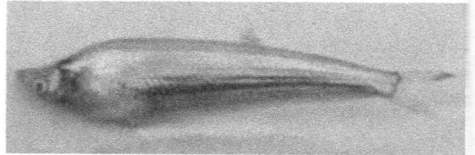

B. 12; D. 4; P. 14-15 (1/13-14); V. 8; A. 60-75 (2-3/58-72); C. 17-18; Barbels 4.

| Local names | : | Puffa |
|---|---|---|
| Environment | : | Demersal, freshwater, brackish, pH range 6.0-8.0; dH range 4.0-28.0; depth range 0.2. |
| Climate | : | Tropical 20-26°C |
| Dangerous | : | Harmless |

**Characters**

1. Length of head ranges from 5 to 7 and the height of body from 5 to 6 in the total length.

2. Body is generally compressed and the head is depressed.

3. Eyes are subcutaneous and situated behind and opposite the angle of the mouth.

4. Barbels four or two in number.

5. Dorsal fin is short, without spine and originates in advance of pelvic origin.

6. Pectorals are with a moderately strong spine, which may be serrated or entire.

7. Anal fin is very long, not continuous with the caudal fin.

8. The caudal fin is forked with its upper lobe slightly longer.

9. Body is silvery, shot with purple. A black spot is present on the shoulder and often one or two faint black lateral bands are found in the upper part of the body.

| Remarks | : | It attains at least 45 cm in length. It is an excellent food fish. |
|---|---|---|
| Importance | : | Fisheries : commercial; aquaculture : commercial; aquarium : commercial. |

### 58. *Wallago attu* (Bloch & Schneidar) 1801

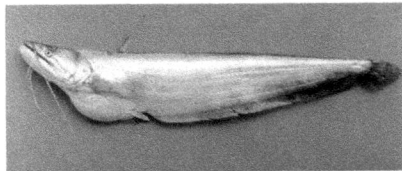

B. 19-21; D. 5; P. 14-16 (1/13-15); V. 8-10; A. 86-93 (4/82-89); C. 17; Barbels 4.

| Local names | : Boalee, Parhin, Shivada, pattan, pari, purrum. |
| Environment | : Demersal, freshwater, brackish |
| Climate | : Tropical 22-25°C; 38° N-10°S |
| Dangerous | : Traumatogenic |

**Characters**

1. Length of head ranges from 5 to 5.5 and the height of body from 6.3 to 6.5 in the total length.

2. Body is compressed and the head is depressed, mouth is oblique.

3. Cleft of mouth extends to nearly an eye diameter behind orbit.

4. Lower jaw is more prominent.

5. Barbels are four in number; maxillary extends beyond anal origin.

6. Dorsal fin short, spineless and as long as pectoral.

7. Pectoral spine is moderately strong and finely serrated.

8. Caudal fin is forked and not united with anal.

9. Body is uniform silvery gray becoming lighter below.

| Remarks | : It attains about 240 cm in length and is often referred to as a freshwater shark on account of its large mouth, this species is good for eating and saltswal. |
| Importance | : Fisheries: commercial; game fish : yes. |
| Order | : Beloniformes. |
| Suborder | : Scomberscoidei. |
| Family | : Belonidae (Needle fishes). |
| Genus | : Xenentodon (Regan) 1911. |

Synopsis of the species of the genus Xenentodon Regan dealt within this text.

## 59. *Xenentodon cancila* (Ham.) 1822

B. 19-21; D. 5; P. 14-16 (1/13-15); V. 8-10; A. 86-93 (4/82-89); C. 17; Barbels 4.

| Local names | : Kawa |
|---|---|
| Environment | : Pelagic, freshwater, brackish, marine; pH range 7.0-7.5; dH range 20.0. |
| Climate | : Tropical 22-28°C; 38°N-5°N |
| Dangerous | : Traumatogenic |

**Characters**

1. Length of head ranges from 2.6 to 2.8 and the height of body from 8 to 12 cm in the total length.

2. Body is elongated, subcylindrical or compressed.

3. Eyes are lateral in position.

4. Both the jaws are prolonged to form a beak.

5. Lower jaw is slightly longer than the upper jaw.

6. Teeth villiform.

7. Dorsal fin originates opposite the anal.

8. Lateral line is not keeled.

9. Scales are small and irregularly arranged.

10. Body is subcylindrical, greenish gray above and whitish below, silvery streak with dark margin extends along the body from opposite.

| Remarks | : It grows upto 40.0 cm in length. This fish is meaty and good for eating especially when lightly stewed or steamed with potherbs. |
|---|---|
| Importance | : Fisheries: minor commercial; aquarium: potential. |
| Order | : Mugiliformes |
| Suborder | : Mugiloidei |
| Family | : Mugilidae (Mullets) |
| Genus | : Mugil Linnaeus 1758 |

Synopsis of the species of the genus Mugil Linnaeus dealt within this text.

## 60. *Mugil cephalus* (Linnaeus) 1758

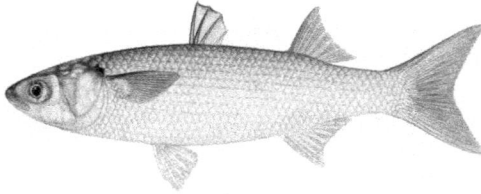

B. 6; D. 4/9 (1/8); P. 15; V. 6(1/5); A. 11 (3/8); C. 15; L.I. 42-44; L.tr. 14.

Popular name      : Grey Mullet

Local names       : Bhomat, Boi

Environment       : Benthopelagic, catamodromous, freshwater, brackish, marine, depth range 0-120cm.

Climate           : Subtropical 8-24°C; 38°N- 42°N.

Dangerous         : Harmless

**Characters**

1. Length of head ranges from 4.5 to 4.7 and the height of body from 5.3 to 5.6 in the total length.

2. Body is oblong and subcylindrical eyes are with or without an adipose lid.

3. Dorsal profile from the snout to the base of the dorsal fin is nearly horizontal.

4. First dorsal arises midway between the end of the snout and the base of the caudal fin.

5. Pectorals are situated above the middle of the depth of the body

6. Anal originates opposite to the second dorsal fin.

7. Body is grayish along the back and silvery below.

8. A dark line is present along each row of scales in the upper half of the body. Cheeks are golden.

9. Pectorals fin has an oblique, deep blue band across its base.

10. Anal is yellowish with a mark along its center and light edge.

Remarks           : It attains at least 120 cm in length. It constitutes an important fishery in chilka lake.

Importance              : Fisheries: highly commercial; aquaculture :
                          commercial; game fish : yes; aquarium: commercial;
                          bait : occasionally.

Synopsis of the species of the genus Rhinomugil Linnaeus dealt within
this text.

**61. *Rhinomugil corsula* (Ham.) 1822**

B. 6; D. 4|17-8; P. 15; V. 6 (1/5); A. 12 (3/9); C. 15; L.I. 48-52; L.tr. 15.

Popular name        : Grey Mullet

Local names         : Khorsula

Environment         : benthopelagic, catamodromous, freshwater,
                      brackish, marine

Climate             : Subtropical

Dangerous           : Harmless

**Characters**

1. Length of head ranges from 4.5 to 4.7 and the height of body from
   6 to 6.5 in the total length.

2. Body stout, head concave between eyes.

3. Eyes projecting above the surface of water.

4. Mouth ventral, protrusible.

5. Dorsal nearer caudal fin then snout tip.

6. Caudal slightly emarginate.

7. Scales finely ctenoid, with a slightly raised line along the center of
   each.

8. Colours of body are dull brown superiorly, becoming lighter along
   the abdomen, dorsal and caudal fins stained with gray.

Remarks              : It attains 45 cm in length and is excellent
                       eating.

| | | |
|---|---|---|
| Importance | : | Fisheries : commercial; aquaculture : commercial. |
| Order | : | Channiformeis |
| Family | : | Channidae |
| Genus | : | Channa Gronovius 1763 |

Synopsis of the species of the genus genus Channa Gronovius dealt within this text.

## 62. *Channa gachua* (Ham.) 1822

B. 5; D. 32-37; P. 15-16; V. 6; A. 21-23; C. 12; L.I. 140-145; L.tr. 3-4 | 6-7.

| | | |
|---|---|---|
| Popular name | : | Brown snake head |
| Local names | : | Dheridhok, Dhok |
| Environment | : | benthopelagic, freshwater, brackish, pH range 6.0-7.0; dH range 15.0. |
| Climate | : | Tropical |
| Dangerous | : | Harmless |

**Characters**

1. Length of head ranges from 3.5 to 4.2 and the height of body from 6 in the total length.

2. Body is elongated, subcylindrical anteriorly and compressed posteriorly.

3. Mouth is large and protractile.

4. Head resembles to that of snakes having large shield like scales above.

5. Eyes are lateral in position.

6. Head scales are broad and irregular.

7. Pectorals fins are as long as head behind the eye.

8. Predorsals are 12 in number, lateral line curves downwards after 12th scale.

9. Body is generally greenish lighter beneath.

10. Dorsal, anal and caudal fins are slaty gray with their margins orange.

11. Pectorals have a deep blue base, transversely barred with orange and blue strips.

Remarks            : It grows upto 20 cm in length. It builds its nest in sheltered crevices in the bank.

Importance         : Aquarium: commercial, Fisheries: Commercial.

**63. *Channa marulius* (Ham.) 1822**

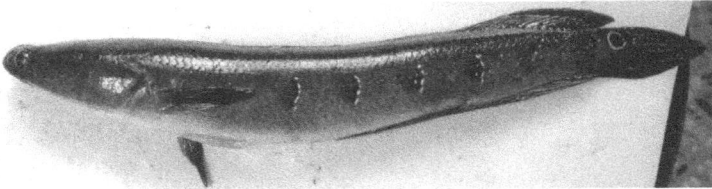

B. 5; D. 45-55; P. 18; V. 6; A. 28-36; C. 14; L.I. 60-70; L.tr. 4½-6½ I 11-13.

Popular name       : Giant snak-heed

Local names        : Pu-muri

Environment        : benthopelagic, freshwater, brackish

Climate            : Tropical 24-28°C

Dangerous          : Harmless

**Characters**

1. Length of head ranges from 4 to 5 and the height of body from 7 to 7.5 in the total length.

2. Body is elongated, subcylindrical anteriorly and compressed posteriorly.

3. Mouth is large and protractile.

4. Head resembles to that of snakes having large shield like scales above.

5. Eyes are lateral in position.

6. Pectorals fin is more than half the length of the head but not reaching the origin of anal fin.

7. Scales are of moderate size on the summit of head.

8. Predorsals are 16 in number; lateral line is straight upto the 17ᵗʰ scale.

9. Body is grayish green becoming lighter below.

10. The young specimens are with a brilliant orange band passing from the tip of the snout, over the eyes to the tip of caudal fin.

11. In the forms nearing maturity, there are 4 to 5 round black blotches situated below the lateral line.

12. There is a large black prominent ocellus with orange boundary at the upper part of the base of the caudal fin.

Remarks : It attains as much as 183 cm in length. It is favourite sporting species and highly esteemed as food.

Importance : Fisheries: commercial; aquaculture: commercial; game fish : yes; aquarium: commercial.

## 64. *Channa punctata* (Bloch) 1793

B. 5; D. 29-32; P. 16-17; V. 6; A. 21-23; C. 12; L.I. 35-40; L.tr. 4-5 | 6-9.

Popular name : Green snake head

Local names : Phool, Dhok

Environment : benthopelagic, freshwater

Climate : Tropical

Dangerous : Harmless

**Characters**

1. Length of head ranges from 3.3 to 3.6 and the height of body from 5.5 to 7 in the total length.

2. Body is elongated, subcylindrical anteriorly and compressed posteriorly.

3. Mouth is large and protractile.

4. Head resembles to that of snakes having large shield like scales above.

5. Eyes are lateral in position.

6. Lower jaw is a longer and 3-6 canine behind a row of villiform teeth.

7. Predorsals are 12 in number; lateral line is practically straight.

8. The scales are large and irregular on the summit of the head.

9. Caudal fin is rounded.

10. Body is generally greenish gray becoming yellow below, several bands pass from the dorsum of the body downwards to the middle of the sides fins are spotted.

Remarks            : It attains about 30 cm in length. This fish is a prolific breeder and its development is rapid.

Importance         : Fisheries: commercial.

## 65. *Channa striatus* (Bloch) 1793

B. 5; D. 37-45; P. 16-18; V. 6; A. 23-26; C. 13; L.I. 50-60; L.tr. 4½-8 | 7-10.

Popular name       : Striped snake head.

Local names        : Maral, sohr, dakhu, mural, morrul.

Environment        : benthopelagic, freshwater, brackish, pH range 7.0-8.0; dH range 20.0; depth range 10 m.

Climate            : Tropical 23-27°C; 35°N-18°S.

Dangerous          : Potential pest.

## Characters

1. Length of head ranges from 3.8 to 4 and the height of body from 6 to 8 in the total length.

2. Body is elongated, subcylindrical anteriorly and compressed posteriorly.

3. Mouth is large and protractile.

4. Head resembles to that of snakes having large shield like scales above.

5. Eyes are lateral in position.

6. Lower jaw is the longer; maxilla reaches to below the hind broder of the eye.

7. The pectoral does not quite reach to above the origin of the anal.

8. Predorsal scales are 18-20 in number, lateral line curves downwards below 12$^{th}$ dorsal ray.

9. Body is grayish to black above, dirty white below.

10. Bands of gray or black pass down from the sides to the abdomen fins are grayish.

11. Young passes a large black ocellus at the end of the base of dorsal fin.

| | | |
|---|---|---|
| Remarks | : | It grows upto 100 cm in length. It prefers stagnant mudy waters and is carnivorous in habit. |
| Importance | : | Fisheries: highly commercial; aquaculture: commercial; aquarium: show aquarium. |
| Order | : | Perciformes |
| Sub-order | : | Anabantoidei |
| Family | : | Anabantidae (Climbing perches, Builders). |
| Genus | : | Anabus Cuvier and Cloquet 1816. |

Synopsis of the species of the genus Anabus Cuvier and Cloquet dealt within this text.

### 66. *Anabas testudineus* (Bloch) 1792

B. 6; D. 17-18 I 8-10; P. 15; V. 6; A. 9-10 I 9-11; C. 17; L.I. 28-32; L.tr. 3-4 I 9/10.

| | | |
|---|---|---|
| Popular name | : | Climbing perch |
| Local names | : | Persh |
| Environment | : | Demersal, freshwater, brackish. |
| Climate | : | Tropical 22-30°C; 28°N-10°S. |
| Dangerous | : | Harmless |

**Characters**

1. Length of head ranges from 3.5 to 3.6 and the height of body from 3 to 4 in the total length.

2. Body is elongated, subcylindrical, compressed or oblong.

3. Accessory respiratory organs are well developed.

4. Mouth is relatively small.

5. Dorsal fin is single and longer than the anal fin; its spinous portion is longer than the soft portion, with 17-18 spines.

6. Anal fin is with 9-10 spines.

7. Scales are ctenoid and sometimes found over the base of the dorsal, pectoral, anal and caudal fins.

8. Lateral line is interrupted.

9. Body is light to dark green above and greenish yellow to orange bellow, four wide cross bands on the body.

10. A black spot is found at the posterior end of the opercule on both the sides.

| | | |
|---|---|---|
| Remarks | : | It grows up to about 26 cm in length. It has a very good flavour and is popular as food. |
| Importance | : | Fisheries : commercial; aquaculture: commercial; aquarium : commercial. |
| Suborder | : | Gobiodei |
| Family | : | Gobiidae (Gobies) |
| Genus | : | Glassogobius Gill 1859 |

Synopsis of the species of the genus Glassogobius Gill dealt within this text.

## 67. *Glossogobius giuris* (Ham.) 1822

B. 4; D. 6|1/8-9; P. 20; V. 6 (1/5); A. 9-10 (1/8-9); C. 17; L.I. 30-35; L.tr. 8-12.

| | |
|---|---|
| Popular name | : Bar-eyed Goby |
| Local names | : Tank Goby |
| Environment | : demersal, amphiodromous, freshwater, brackish, marine. |
| Climate | : Tropical 24-45°C; 32°S |
| Dangerous | : Harmless |

**Characters**

1. Length of head ranges from 3.5 to 4 and the height of body from 5 to 5.5 in the total length.

2. Body is generally elongated; mouth is usually larger with small canine teeth.

3. Head is depressed, pointed and the lower jaw is the longer.

4. Scales extend over the head; some are present on the upper part of cheeks and opercles. Body scales are caninoid while the head scales are cycloid.

5. Dorsal fins are low in number, the first dorsal with 6 week spines.

6. Anal is pointed posteriorly and with one weak spine.

7. The pelvic fins have become united at their bases to forma sucking disc.

8. Caudal fin is oblong; lateral line scales are 30-35.

9. Body is with variable colours generally olive to dusky green above, lighter below with light black markings on the head, irregular markings on the body are also present fins are yellow green with spots and dark markings.

Remarks                    : It grows up to a 50 cm in length. It is very
                             popular food fish.

Imporatnce                 : Fisheries: minor commercial; aquaculture:
                             commercial; aquarium: commercial. This fish
                             is highly esteemed as food. Bele is one of the
                             varieties found in both fresh and brackish water,
                             largely caught and eaten. Good sport on rod and
                             line with a bait of small prawn.

Suborder                   : Percoidei

Family                     : Ambassidae (Glassy Perchlets)

Genus                      : Chanda Hamilton 1822

Synopsis of the species of the genus Chanda Hamilton dealt within
this text.

### 68. *Chanda nama* (Hamilton) 1822

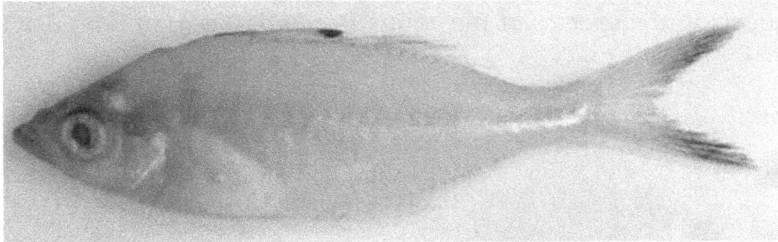

B. 6; D. 1+7|14-18(1/13-17); P. 12-13; V. 6; A. 17-21(3/14-18); C. 17.

Popular name       : Indian glass fish.

Local names        : Kachki

Environment        : Benthopelagic, freshwater, brackish.

Climate            : Tropical 38°N-6°N.

Dangerous          : Harmless

### Characters

1. Length of head ranges from 4 to 4.5 and the height of body from
   2.7 to 3.5 in the total length.

2. Body is generally compressed and both the profile of the body is
   equally convex.

3. Lower jaw is much longer than the upper jaw.

4. Preorbital is slightly is denticulated.

5. Dorsal fins are low, both are united at their bases, first dorsal is with 7 spines; a recumbent spine is also present before it. Second dorsal spine is the longest.

6. Anal fin is with spines; scales are minute.

7. Lateral line may be broken absent or even indistinct.

8. Body is whitish yellow with black dots all over the body.

9. The dots on the shoulder form an oblong vertical path.

10. Summit of the head and the top of eyes are black.

Remarks            : It grows upto abut 11.0 cm in length. It is a favorite aquarium fish and it is called as glassfish or X-ray fish.

Importance         : Fisheries : minor commercial; aquarium : show aquarium.

Synopsis of the species of the genus *Parambassis* Hamilton dealt within this text.

## 69. *Parambassis ranga* (Ham.) 1822

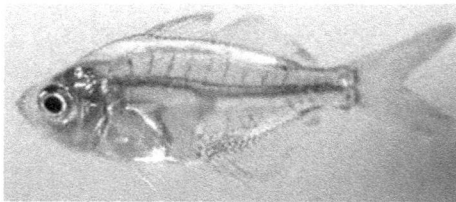

B. 6; D. 1+7∣14-15(1/13-15); P. 12-13; V. 6 (1/5); A. 17-19 (3/14-16); C. 17.

Popular name       : Indian glass fish.

Local names        : Chamberdi, Kachki.

Environment        : demersal, freshwater, brackish.

Climate            : Tropical 20-30°C; 38°N-1°S.

Dangerous          : Harmless

## Characters

1. Length of head ranges from 3.2 to 4 and the height of body from 2.3 to 2.7 in the total length.

2. Both the profiles of the body are very convex but the profile over the eyes is slightly concave.

3. The double horizontal margin of the preopercle is serrated but the inter and subopercles are entire.

4. Dorsal fins are two in number, the first with 7 spines.

5. Anal fin is with 3 spines.

6. Caudal fin is forked.

7. Teeth are villiform and present on jaws, vomer, and palate and sometimes on the tongue.

8. Body is white with black spots closely set dots form a dark mark on the shoulder of each side.

9. Margins of vertical fins are dark or gray.

Remarks               : It grows upto 10 cm in length. It is also good favourite aquarium fish.

Importance            : Fisheries : subsistence fisheries; aquarium : commercial

Family                : Cichlidae (Chromides or Cichlids)

Genus                 : Oreochromis Smith, 1849

Synopsis of the species of the genus Oreochromis Smith dealt within this text.

### 70. *Oreochromis mossambicus* (Peters, 1852)

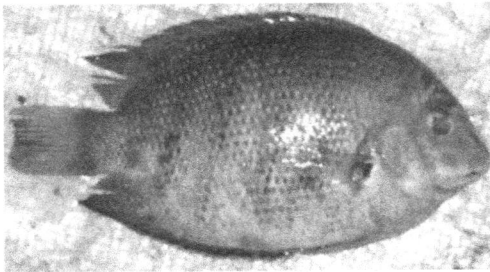

B. 6; D. 15+17|10-13; P. 15; V. 6(1/5); A. 3|7-12; C. 16; L.I. 18-21; L.tr. 10-15.

Popular name          : Thilapia, Tilapi

Local names           : Tilapi

Environment           : Benthopelagic, amphiodromus, freshwater, brackish, depth range 10 m.

Climate               : Tropical 8-24°C; 13°S-35°S.

Dangerous             : Potential pest.

**Characters**

1. Body short, more or less elongate abdomen rounded.

2. Head compressed, with concave upper profile.

3. Mouth terminal, large at least width of head or often nearly as wide head.

4. Dorsal fin inserted above base of pectoral, with 15 or 16 spines and 10 or 11 rays.

5. Anal fin with 3 spines (rarely four), third spines a little longer than the dorsal.

6. Caudal fin rounded, may be truncate in the young.

7. Scales are cycloid.

8. Lateral line incomplete, upper one with 18-21 and lower one with 10-15 scales.

Remarks               : It grows upto 39.0 cm in length. The fish is reported to be unsuitable for culture along with major Indian carps, because of its depredation on carp fry.

Importance            : Fisheries : highly commercial; aquaculture : commercial; game fish : yes; aquarium : commercial.

Family                : Nandidae (leaf fishes).

Genus                 : Badis Bleeker 1853.

Synopsis of the species of the genus Badis Bleeker dealt within this text.

**71. Badis badis (Ham.) 1822**

B. 6; D. 16-18|7-10; P. 12; V. 6 (1/5); A. 9-11 (3/6-8); C. 16; L.I. 26-32; L.tr. 2½|8½.

| | | |
|---|---|---|
| Popular name | : | Badis |
| Local names | : | Koi bandi |
| Environment | : | Benthopelagic, freshwater, pH range 6.8; dH range 5-19. |
| Climate | : | Tropical 23-26°C. |
| Dangerous | : | Harmless |

**Characters**

1. Length of head ranges from 4.5 to 3.6 and the height of body from 4 in the total length.

2. Body is oblong and compressed. Head is usually large; mouth is protractile.

3. Oracle has one sharp spine.

4. Villiform teeth are present in the jaws, vomer, palatines and epiphyal.

5. Dorsal fin is single; the soft portion of the fin is rather elevated and pointed.

6. Anal spines are short, ctenoid scales are present.

7. Lateral line is interrupted.

8. Body is dark black when the fish is fresh.

9. A black spot is present behind the oprecle and above the pectoral fin.

10. Another prominent black spot behind the eye and longer black spot is located at the base of caudal fin.

11. Body is spotted with irregular black spots.

| | | |
|---|---|---|
| Remarks | : | It grows upto a length of about 9 cm. |
| Importance | : | Fisheries: of no interest; aquarium: commercial |
| Order | : | Mastacembeliformes. |
| Family | : | Mastacembelidae (spiny Eels). |
| Genus | : | Mastacembelus Gronovius 1763. |

Synopsis of the species of the genus Mastacembelus Gronovius dealt within this text.

## 72. *Mastacembelus armatus* (Lacepede) 1800

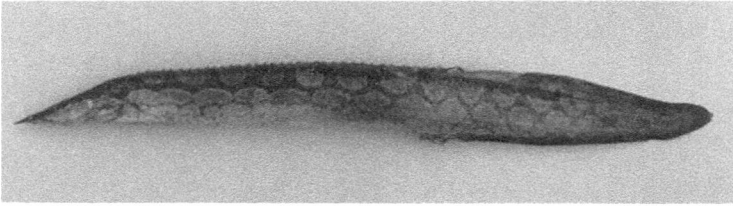

B. 6; D. 32-39 I 74-90; P. 23-25; A. 78-91(3/75-88).

| | |
|---|---|
| Popular name | : Spiny Eel |
| Local names | : Bam, wam. |
| Environment | : demersal, freshwater, brackish, pH range 6.5 – 7.5; dH range 15.0. |
| Climate | : Tropical 22-28°C; 38°N-1°N. |
| Dangerous | : Harmless |

**Characters**

1. Length of head ranges from 5.5 to 7.5 and the height of body from 10.3 to 12 in the total length.
2. Body is elongated and eel-like.
3. The cleft of mouth is narrow; teeth are minute and present in the jaws.
4. The snout is long trilobed and with a fleshy appendage.
5. The dorsal spines commence over the middle of the pectoral fin.
6. Origin of the soft part of the dorsal fin is behind the origin of anal.
7. The anal has three spines; caudal fin is united with the dorsal and anal.
8. The caudal fin is rounded.
9. Pectorals fins are small and the Ventrals fins are absent.
10. Scales are very minute and the head of body scales are similar.
11. Body is brownish above and lighter below.
12. In some forms a faint undulating band is present above the lateral line and rounded black spots along the base of the dorsal fin.

Remarks          : It attains up to 90 cm in length. It is highly esteemed as food.

Importance          : Fisheries : commercial; aquarium : commercial.

Synopsis of the species of the genus Macrognathus Gronovius dealt within this text.

### 73. *Macrognathus pancalus* (Ham.) 1822

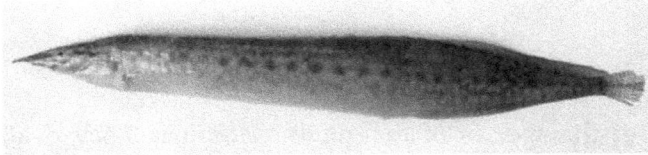

B. 6; D. 24-26|30-42; P. 19; A. 78-91 (3/75-88).

Popular name          : Spiny Eel

Local names          : Wam, Wambat

Environment          : Demersal, freshwater, brackish.

Climate          : Tropical

Dangerous          : Harmless

**Characters**

1. Length of head ranges from 5 to 6 and the height of body from 7.3 to 10 in the total length.

2. Body is elongated, cylinderical and eel-like.

3. Snout is long and trilobed its extermity.

4. Preorbital spines are present.

5. The dorsal spines are short and gradually increasing in length towards the posterior side.

6. Origin of the soft dorsal fin is behind the origin of anal.

7. Caudal fin is not united with the dorsal and anal.

8. Anal is 3 spines, the second spines is longest and strongest.

9. Pectorals are small and the Ventrals fins are absent.

10. Caudal fin is rounded.

11. Body is greenish olive along the back and yellowish beneath, yellowish white spots are present over the side of the body.

Remarks : It attains at least 18cm in length. It is highly esteemed as food.

Importance :

Order : Cyprinodontiformes

Suborder : Cyprinodontoidei

Family : Poecilidae

Genus : Gambusia Poey 1854.

Synopsis of the species of the genus Gambusia Poey dealt within this text.

## 74. *Anguilla bengalensis bengalensis* (Gray, 1831)

Class : Teleostomi

Order : Angguilliformes

Family : Anguillidae

Genus : Anguilla

Species : Bengalensis bengalensis

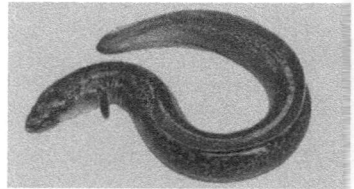

D. 250-305, A. 220 - 250; L.I. 160-180.

Popular names : Indian mottled eel.

Local names : Ahir, Aheer, Eel.

Environment : Marine; freshwater; brackish; benthopelagic; catadromous.

Climate : Tropical; 23N-33 S.

Dangerous : Harmless

## Characters

1. Length of head ranges from 5.5 to 7.5 and the height of body from 10.3 to 12 in the total length.

2. Body is elongated and eel-like.

3. Mouth terminal, lips prominent, narrow bands of teeth on jaws, broad band on vomer.

4. Head is conical flattened dorsally.

5. Origin of the soft part of the dorsal fin is behind the origin of anal.

6. The caudal fin is rounded.

7. Scales are very minute and the head of body scales are similar.

8. Body is brownish above and lighter below.

9. In some forms a faint undulating dark band is present on the body.

Remarks            : Most common eel in Indian inland waters.

Importance         : Fisheries : commercial; aquaculture : commercial; game fish : yes.

Maximum size       : 200 cm TL

## 75. *Gambusia affinis* (Baird and Girard) 1853

D. 6-9; A. 8-10; V. 6; L.I. 32; L.tr. 8.

Popular name       : Mosquito fish

Local names        : Gambusia

Environment        : benthopelagic, potamodromous,freshwater, brackish, pH range 6-8; dH range 15-19.

Climate            : Subtropical 12-29°C; 42°N-26°N.

Dangerous          : Potential pest

## Characters

1. Length of head ranges from 3.6 to 4 and the height of body from 3.5 to 4 in the total length.

2. Body is somewhat elongated.

3. Eye diameter is 3 in the total length of head.

4. In males the dorsal fin is situated in the middle of the body while in females it is situated midway between the front margin of eye and the tip of the caudal fin.

5. Anal fin is with 8-10 fin rays.

6. Lateral line scales are 32.

7. Faint dark lines are present in the upper half of the caudal fin.

8. Sides of the body are irregularly dotted black.

9. Two or three cross bands of dots are present on the dorsal and caudal fins.

| | | |
|---|---|---|
| Remarks | : | It grows upto 4.0 cm in length. It is a small size of fish and is a very good Larvivorus. |
| Importance | : | Fisheries : minor commercial; aquarium : commercial. |
| Subfamily | : | Poeciliinae. |
| Genus | : | Poecilia. |

Synopsis of the species of the genus Poecilia Peters, 1859 dealt within this text.

## 76. *Poecilia reticulata* (Peters) 1859

D. 7-9; A. 8-10; V. 6; L.l. 34; L.tr. 8.

| | | |
|---|---|---|
| Popular name | : | Mosquito fish |
| Local names | : | Guppy |
| Environment | : | benthopelagic, potamodromous,freshwater, brackish. |
| Climate | : | Subtropical |
| Dangerous | : | Potential pest |

## Characters

1. Length of head ranges from 3.7 to 4 and the height of body from 3.6 to 4.2 in the total length.

2. Body is somewhat elongated.

3. This fish is very similar to Gumbusia.

4. In males the length of anal fin rays in more or less equal to length of gonopodium.

5. Generally black spots and other brilliant colouration present sides of the body are irregularly dotted black.

6. Two or three black dots on the caudal and dorsal fin.

Remarks : It grows upto 4.0 cm in length. It is a small size of fish and is a very good Larvivorus. It is used as live food for carnovorous aquarium fish.

Importance : Fisheries : minor commercial; aquarium : commercial.

### 77. *Poecilia formosa* (Girard) 1859

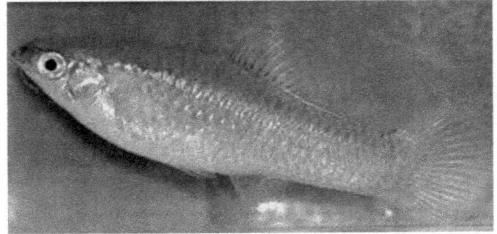

D. 10-12; A. 12-15; V. 6; L.I. 39; L.tr. 9.

Popular name : Mosquito fish

Local names : Guppy

Environment : benthopelagic, non-migratory, freshwater, brackish.

Climate : Subtropical 27°N-25°N.

Dangerous : Harmless

### Characters

1. Body is somewhat elongated.

2. First three anal rays are unbranched.

3. They inhibit fresh and brackish waters and bear the young alive.

5. Body on the brown spots on side (may have row of dusky black spots).

Remarks : It is a very good Larvivorus fish.

Importance : Fisheries : of no interest; aquarium: commercial.

# ABBREVIATION

A       : Anal fin

B       : Branchiostegal rays

C       : Caudal fin

D       : Dorsal fin

O       : Adipose fin

P       : Pectoral fin

V       : Ventral or Pelvic Fin

L.I.     : Lateral line of Perforated scales

L.r.     : Lateral row of unperforated scales

L.tr.    : Lateral transverse rows of scales

FMA   : Freshwater of Marathwada region

# REFERENCES

Bhuiyan A.L. (1964): Fishes of Dacca, *Asiat. Soc. Pakistan*, Pub. 1, No. 13, Dacca, pp. 60-61.

Daniels R. (2002): Freshwater Fishes of Peninsular India. University Press.

Datta M.J.S. and M.P. Shrivastva (1988): Natural History of Fishes and Systematics of Freshwater Fishes of India. Narendra Publishing House, Delhi.

Day F. (1878): The Fishes of India: being a natural history of the fishes known to inhabit the seas and freshwaters of India, Burma and Ceylon. Text and Atlas in 4 Parts. London.

Hamilton F. (1822): An account of the Fishes Found in the River Ganges and its Branches. Edinburgh & London. An account of the fishes found in the river Ganges and its branches.: i-vii + 1-405, Pls. 1-39.

Heda N. (2007): Some Studies on Ecology and Diversity of Freshwater Fishes in the Two Rivers of Vidarbha Region of Maharashtra (India). A Thesis submitted to Sant Gadgebaba Amravati University, Maharashtra (India).

Hiware C.J. (2006): Ichthyofauna from Four Districts of Marathwada Region, Maharashtra, India. *ZOOS' PRINT JOURNAL* 21(1): 2137-2139.

Jadhav B.V, Kharat S.S, Raut R.N, Paingankar M & Dahanukar N. (2011): Freshwater Fish Fauna of Koyna River, northern Western Ghats, Indi. *Journal of Threatened Taxa* 3(1): 1449-145.

Jayaram K. C. (1981): The Freshwater Fishes of India, Pakistan, Bangladesh, Burma and Sri Lanka—A Handbook. The Director, Zoological Survey of India, Kolkata.

Jayaram K. C. (2002): Fundamentals of Fish Taxonomy. Narendra Publishing House, New Delhi. Jayaram K.C. (1999): The Freshwater Fishes of the Indian Region. Narendra Publishing House, New Delhi.

Jayaram, K.C. (2006): The Catfishes of India, Narendra Publishing House, New Delhi, i-xxii+383, Pls. i-xiii.

Menon, A.G.K. (1987): The Fauna of India and Adjacent Countries, Pisces, Vol 4, Teleostei-Cobitoidea, Part 1, Homalopteridae. Zoological Survey of India, Kolkata.

Menon, A.G.K. (1992): The Fauna of India and Adjacent Countries, Pisces, Vol 4, Teleostei-Cobitoidea, Part 2 Cobitidae. Zoological Survey of India, Kolkata.

Menon, A.G.K. (1999): Check list—Freshwater Fishes of India, Rec. Zool. Surv. India, Occ. Paper. No.175: i-xxix, 1-366 pp., Zoological Survey of India, Kolkata, India.

Menon, A.G.K. (2004). Threatened Fishes of India and Their Conservation. Zoological Survey of India, Kolkata.

Nelson, J.S. (2006): Fishes of the World. Fourth Edition, John Wiley & Sons, Inc. 1- 601.

Ponniah, A.G. and Sarkar, U.K. (2000): Fish Biodiversity of North-East India. ISBN 81-901014-1-2, NBFGR.NATP PUBL.2, 1-228 pp. National Bureau of Fish Genetic Resources, Lucknow, U.P-226 002, India.

Qureshi L., T. A. Qureshi, N. A. Qureshi (1983): Indian Fishes: Classification of Indian Teleosts. Publisher Brij Bros., 224 pages.

Sakhare, V.B. (2007): Applied Fisheries, Daya Publishing House, Delhi.

Sakhare, V.B. (2007): Reservoir Fisheries and Limnology, Narenra Publsihng House, Delhi.

Sykes, W.H. (1839): On the fishes of the Deccan. Proceedings of the General Meetings for Scientific Business of the Zoological Society of London 6 : 157-165.

Talwar, P.K. & A.G. Jhingran (1991): Inland Fishes of India and Adjacent Countries. Oxford-IBH Publishing Co. Pvt. Ltd., New Delhi.

www.ingramcontent.com/pod-product-compliance
Lightning Source LLC
Chambersburg PA
CBHW021433180326
41458CB00001B/261